SPRINGER TRACTS
IN MODERN PHYSICS

Ergebnisse
der exakten Natur-
wissenschaften

Volume **39**

Springer-Verlag Berlin Heidelberg GmbH 1965

Herausgeber:

G. HÖHLER, Institut für Theoretische Kernphysik der Technischen Hochschule, 75 Karlsruhe, Hauptstraße 54

Schriftleiter:

E. A. NIEKISCH, Kernforschungsanlage Jülich, Arbeitsgruppe Institut für Technische Physik, 517 Jülich, Postfach 365

ISBN 978-3-662-15903-3   ISBN 978-3-540-37142-7 (eBook)
DOI 10.1007/978-3-540-37142-7

Brühlsche Universitätsdruckerei Gießen

Electron and Photon Interactions at High Energies

Invited Papers
Presented at the International Symposium
Hamburg, June 8—12, 1965

Contents

The research contributions, the closing remarks by Prof. ROBERT R. WILSON and the discussion following the invited papers will be contained in the forthcoming "Proceedings of the International Symposium on Electron and Photon Interactions at High Energies, Vol. I and II".

Theory of Strong Interactions of Elementary Particles in the GeV Region

L. Van Hove

Theory Division
CERN, Geneva (Switzerland)

This lecture attempts to review the main aspects of strong interaction processes in the high energy region where cross-sections show a smooth energy dependence suggestive of an asymptotic behaviour. Due to a steady increase of accurate and reliable experimental data the general properties of this asymptotic region manifest themselves more and more clearly. In absence of a true theory, the theoretical analysis is essentially of a phenomenological character. Either it correlates data on the basis of a few general principles and assumptions, or it fits them by means of tentative theoretical models. Despite its modest scope, this type of analysis appears to be quite successful in developing gradually a general picture of strong interactions at high energies.

I. Elastic Scattering at Small Momentum Transfer

We discuss first elastic scattering processes $A + B \rightarrow A + B$ (A, B are hadrons) in the region of c.m. momentum transfers $\Delta = \sqrt{-t}$ defined by

$$0 < -t \lesssim 1 \,(\text{GeV}/c)^2 \,. \tag{I.1}$$

Good data are available for a number of reactions, and the general features appear to be the same for all. For laboratory momentum $p_L \gtrsim 6 \,\text{GeV}/c$ a good description is obtained by means of an asymptotic series expansion of the scattering amplitude containing two or three terms. We give it the general form

$$T(st) \underset{s \rightarrow +\infty}{\cong} \sum_{j=1,2,\dots} c_j(t)\, s_t^{\alpha_j(t)} \left(\ln s_t - \frac{i\pi}{2} \right)^{\beta_j(t)} \tag{I.2}$$

s is the square of the c.m. energy, and for later convenience the expansion is taken to be in decreasing powers of the quantity

$$s_t = s + \frac{t}{2} - m_A^2 - m_B^2 \,. \tag{I.3}$$

$\alpha_j(t)$, $\beta_j(t)$ are real exponents, $c_j(t)$ can be complex. One has $\alpha_1(t) >$ $> \alpha_2(t) > \cdots$.

(I.2) is supposed to represent the spin-independent part of the elastic scattering amplitude. Spin-dependent terms of the amplitude can be expanded in the same way, but they are generally expected to grow less rapidly with s than the spin-independent term*.

One could add in each term of (I.2) factors having a variation slower than logarithmic, for example $(\ln \ln s_t)^{\gamma_j(t)}$. The data available, however, are much too limited to decide on the presence or absence of such factors. In fact it is even impossible to establish or reject the presence of the factors $(\ln s_t)^{\beta_j(t)}$ in (I.2). The analysis is therefore usually carried out by putting $\beta_j(t) = 0$.

The expansion (I.2) is general enough to accommodate various types of singularities of the scattering amplitude in the complex angular momentum plane of the t channel. Branch points give $\beta_j(t) \neq 0$, poles $\beta_j(t) = 0$. The poles can be moving (Regge poles) or fixed depending on whether $\alpha_j(t)$ depends on t or not. As mentioned above experiment does not allow to decide between the occurrence of poles or branch points. It is now generally recognized that both types of singularities are theoretically possible. In fact, branch points are expected to be present[1].

Using analyticity of T in s and crossing symmetry**, the expansion (I.2) can be related to the corresponding asymptotic expansion for the crossed reaction $\bar{A} + B \to \bar{A} + B$

$$\bar{T}(ut) \underset{u \to +\infty}{\cong} \sum_{j=1,2,\ldots} c_j^*(t)\, e^{-i\pi\alpha_j(t)}\, u_t^{\alpha_j(t)} \left(\ln u_t - \frac{i\pi}{2}\right)^{\beta_j(t)} \qquad (I.4)$$

where u is the square of the c.m. energy and

$$u_t = u + \frac{t}{2} - m_A^2 - m_B^2. \qquad (I.5)$$

The coefficients $c_j(t)$, $\alpha_j(t)$, $\beta_j(t)$ are the same as in (I.2). The variables s_t, u_t in (I.2) and (I.4) have been so chosen as to give this very simple form to the high energy consequences of analyticity and crossing symmetry***.

All data on direct and crossed reactions are consistent with (I.2) and (I.4). It is in fact very useful to use these equations simultaneously in order to determine the parameters $c_j(t)$, $\alpha_j(t)$ as accurately as possible. We recall that for the time being $\beta_j(t)$ cannot be determined from the data and is usually put equal to zero, so that (I.2) and (I.4) reduce to

$$T(st) \underset{s \to +\infty}{\cong} \sum_{j=1,2,\ldots} c_j(t)\, s_t^{\alpha_j(t)} \qquad (I.6)$$

$$\bar{T}(ut) \underset{u \to +\infty}{\cong} \sum_{j=1,2,\ldots} c_j^*(t)\, e^{-i\pi\alpha_j(t)}\, u_t^{\alpha_j(t)}. \qquad (I.7)$$

---

* Experimental confirmation of this conjecture would be highly desirable. For πp collisions it would require polarization and spin correlation measurements by means of polarized proton targets. For pp collisions such measurements should be done with incident and target protons being both polarized.

** One is now very close to being able to derive the necessary properties from axiomatic field theory [2].

*** For a proof, see [3].

1. The forward scattering amplitude. Our knowledge of the forward scattering amplitude $T(s0)$ stems from the data on total cross-sections σ_T, related to $T(s0)$ by the optical theorem

$$\sigma_T = \left(8\pi^2/ks^{\frac{1}{2}}\right) \operatorname{Im} T(s0), \quad k = \text{c.m. momentum} \tag{I.8}$$

and from the recently acquired data on the ratio

$$X = \frac{\operatorname{Re} T(s0)}{\operatorname{Im} T(s0)} . \tag{I.9}$$

The by now familiar total cross-section curves are reproduced in Fig. 1 [4]. The ratio X is obtained from elastic scattering measurements at very small $|t|$ where the nuclear and Coulomb amplitudes interfere. One assumes the scattering to be spin-independent and one assumes $T(st)$

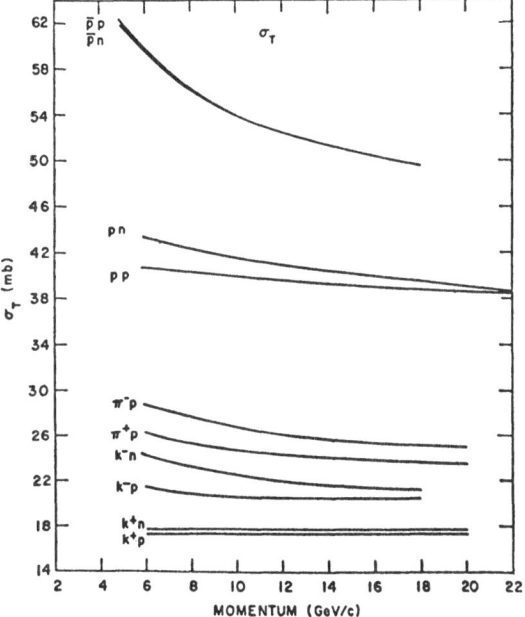

Fig. 1. Total cross-sections for nucleon-nucleon and meson-nucleon scattering, reproduced from Ref. [4]

to be constant between $t = 0$ and the small $|t|$ where the interference is measured. The results available to date are given in Figs. 2 and 3 [5, 6]. There appears to be compatibility with dispersion relations for pp scattering. The situation is less clear for πp where the data for forward charge exchange scattering [7, 8] are also to be taken into account.

We discuss the general behaviour of σ_T and X in terms of (I.6—7). The total cross-section curves indicate with good accuracy

$$\alpha_1(0) = 1 . \tag{I.10}$$

Keeping only the leading term with $j = 1$, (I.6) and (I.10) then give

$$\operatorname{Im} T(s0) \cong \operatorname{Im} \bar{T}(s0), \quad \operatorname{Re} T(s0) \cong -\operatorname{Re} \bar{T}(s0) \qquad (I.11)$$

for $s \to +\infty$. The first relation is the Pomeranchuk theorem $\sigma_T \cong \bar{\sigma}_T$ for total cross-sections [9]. Its asymptotic validity is compatible with the data, although up to 30 GeV there are considerable differences $\sigma_T - \bar{\sigma}_T$ which, for $p_L \gtrsim 6$ GeV/c, can be well described by the second term $j = 2$ of (I.6) and (I.10), $\alpha_2(0)$ being of order 0.5.

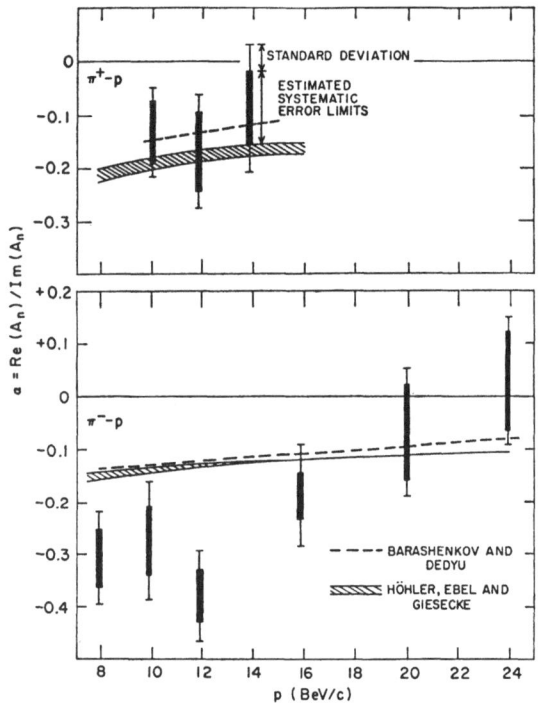

Fig. 2. The ratio X of real to imaginary parts of the forward scattering amplitude for pion-proton scattering, following Ref. [5]

The second relation (I.11) gives $X \cong -\bar{X}$ for $s \to +\infty$, where $\bar{X}$ is the analogue of (I.9) for the crossed reaction. Experiment shows X and $\bar{X}$ to be negative where measured, so that (I.11) suggests

$$X \to 0 \quad \text{and} \quad \bar{X} \to 0 \quad \text{for} \quad s \to \infty. \qquad (I.12)$$

This would mean that the coefficient $c_1(0)$ in (I.6) is purely imaginary

$$c_1(0) = i\,|c_1(0)|. \qquad (I.13)$$

The experimental results are entirely compatible with this expectation which implies that the measured values of X are due to the $j = 2$ term of the expansions (I.6—7), with $\alpha_2(0) \cong 0.5$ as before. It should be noted, however, that there would be no theoretical objection against X becoming positive for some reactions in the region $p_L > 30 \,\mathrm{GeV}/c$, if σ_T would show a minimum in this region*.

A detailed analysis of σ_T and X for pp and $\overline{\mathrm{p}}\mathrm{p}$ has been carried out by A. BIAŁAS and E. BIAŁAS on the basis of expansions of type (I.6—7) [11]. It reveals that the high energy limit of (I.9) for pp is compatible

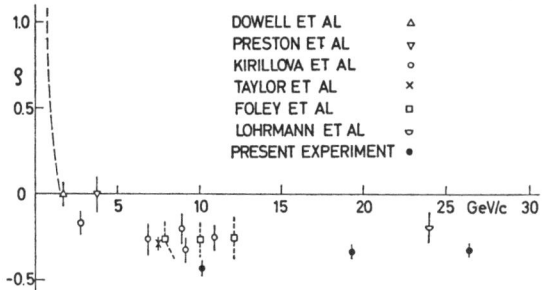

Fig. 3. The ratio X for proton-proton scattering from various experiments, as given in Ref. [6]

with 0 and is very likely to be contained between -0.2 and 0. Furthermore it shows that measurement of (I.9) for $\overline{\mathrm{p}}\mathrm{p}$ at laboratory momenta above $5 \,\mathrm{GeV}/c$ is by far the best way of reaching a better determination of the expansion parameters $c_j(0)$, $\alpha_j(0)$ for the proton-proton system.

2. The shape of the diffraction peak. All diffraction peaks measured until now have in the momentum transfer interval (I.1) a common shape given by

$$\frac{\mathrm{d}\sigma}{\mathrm{d}t} = \left(\frac{\mathrm{d}\sigma}{\mathrm{d}t}\right)_{t=0} e^{at+bt^2} . \tag{I.14}$$

The coefficient a, although depending somewhat on the reaction and the energy, is always of order $10 \,(\mathrm{GeV}/c)^{-2}$. The dimensionless quantity b/a^2 is not well determined, but it is always positive and has an order of magnitude

$$\frac{b}{a^2} \cong 0.01 \quad \text{to} \quad 0.04 . \tag{I.15}$$

From (I.6—7) taken at energies sufficiently high for the leading term $j = 1$ to be the only important one, one has $|T|^2 \cong |\overline{T}|^2$ and $\dfrac{\mathrm{d}\sigma}{\mathrm{d}t}$ therefore becomes the same for the direct and the crossed reaction:

$$\frac{\mathrm{d}\sigma}{\mathrm{d}t} \cong \frac{\mathrm{d}\overline{\sigma}}{\mathrm{d}t} \tag{I.16}$$

where both sides are taken at the same energy, $s = u \to +\infty$. (I.16) is the generalized Pomeranchuk theorem [12]. It implies

$$a \cong \overline{a}, \quad b \cong \overline{b} \tag{I.17}$$

at very high energy (the bar always refers to the crossed reaction).

---

* The author thanks Professors V. S. BARASHENKOV, and G. HÖHLER for a discussion of this point. Qualitative arguments tending to explain the negative sign of X, as given for example in H. E. CONZETT [10], are not compelling because the sign of X can always be modified by a real optical potential. In fact, as shown by A. BIAŁAS, and E. BIAŁAS [11], the negative sign of X is related to the fact that σ_T and $\overline{\sigma}_T$ decrease toward their asymptotic limit; an increase would make X positive.

(I.17) is in good agreement with experiment for $\pi^{\pm} p$ scattering for $p_L > 6$ GeV/c. Both a and b appear to be constant with energy (this result is only significant for a, the precision on b being still poor). For pp and $\bar{p}p$ scattering, there is a pronounced energy variation of a (referring to pp), which increases, as well as of $\bar{a}$ (referring to $\bar{p}p$) which decreases. As shown by Czyzewski et al. [13], from a compilation of all data, a and $\bar{a}$ appear to tend to a common limit $\cong 9 \, (\text{GeV}/c)^{-2}$, about the same value as obtained for $\pi^{\pm} p$. This is clearly indicated by Fig. 4, taken from Ref. [13], where our quantity a is denoted by A.

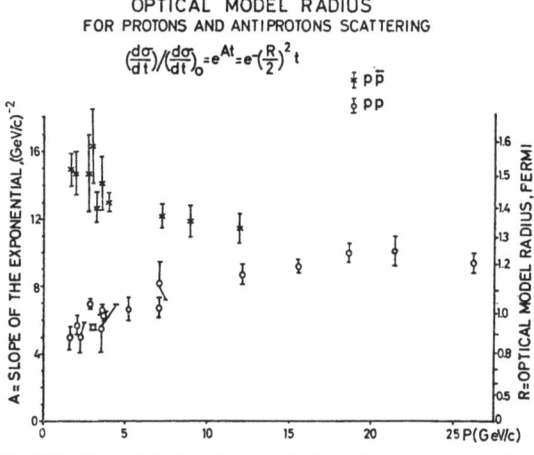

Fig. 4. The width of the diffraction peak for $\bar{p}p$ and pp as a function of laboratory momentum, following Ref. [13]

Therefore, in all cases mentioned, we conclude that the leading term of the expansions (I.6), (I.7) can be given the exponent

$$\alpha_1(t) = 1 \quad \text{for} \quad 0 < -t < 1 \, (\text{GeV}/c)^2 . \tag{I.18}$$

From general considerations [14] one is then led to expect

$$c_1(t) = i \, |c_1(t)| = i \, e^{\frac{1}{2}(at + bt^2)} \tag{I.19}$$

for all t values in the same interval, thereby extending (I.13). The results (I.18–19) for the leading term of (I.6–7) correspond to the familiar diffraction picture characterized by constant cross-sections and purely imaginary amplitude. Deviations thereof will be produced by the higher terms of the expansion; the data roughly suggest $\alpha_2(t) \cong 0.5$ for the t interval considered.

Although our conclusion is that the diffraction picture provides by far the most natural interpretation of the experimental facts, the latter are also compatible with a Regge pole model using several Regge trajectories with positive slope. When several trajectories contribute, the shrinkage effects resulting from their positive slope can be cancelled over a finite energy interval by expansion (antishrinkage) effects produced through appropriate compensations between the various trajectories. For example, a Regge pole fitting of this sort has been made recently by PHILLIPS and RARITA for πN, KN and $\overline{K}$N scattering, using three trajectories for πN and five for KN, K$\overline{N}$ [15]. The final decision between diffraction picture and Regge pole model will probably require data at much higher energies than now available.

II. Dominant Inelastic Collisions

Our knowledge of the most common inelastic collisions, i.e., those collisions which make up most of the inelastic cross-section, is still very limited, both theoretically and experimentally. We limit our discussion to two aspects of the theoretical work devoted recently to this problem.

1. Peripheral, multiperipheral and centroperipheral models. Because of the striking property of the dominant inelastic collisions to have low transverse momenta for all outgoing particles, one has tried for many years to account for them in terms of one meson exchange diagrams.

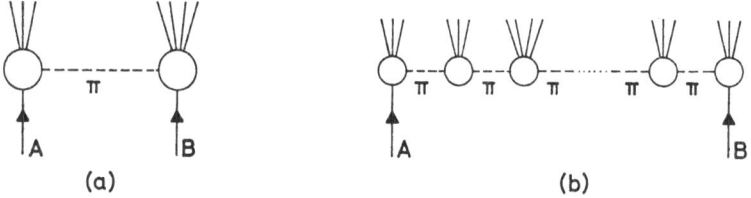

(a)   (b)

Fig. 5a and b. The diagrams characterizing the peripheral model in its original (a) and extended forms (b)

This so-called peripheral model, originally based on diagram (a) of Fig. 5, was soon realized to be inconsistent with itself because, for increasing energy of the collision AB, at least one of the collisions πA, πB in diagram (a) has an increasing energy and should itself become peripheral. Self-consistency is re-established by considering the sum of all diagrams of type (b) with $1, 2, 3 \ldots$ vertices, and specifying some restrictive property for the vertex which prevents it from becoming itself a peripheral collision. Two ways of doing this have been proposed.

The first procedure is to say that the energy of each vertex in its own frame remains below a certain maximum. This gives the multiperipheral model of AMATI, STANGHELLINI and FUBINI [16]. The resulting shadow scattering at high energy is similar to the elastic scattering produced by a single Regge trajectory with positive slope and is therefore not compatible with the diffraction picture.

In the second procedure, recently proposed by DREMIN et al. [17]*, each vertex is regarded to represent a central collision. To this effect a specific algebraic form is assumed for the shadow scattering amplitude resulting from a single vertex, i. e., from central collisions, giving it the form $sf(t)$ when the two incident particles are on their mass shell. For the shadow scattering of the total amplitude, this gives the diffraction picture at very large energy, with additional contributions of peripheral nature in the second and higher terms of the expansion (I.2). The singularities in the angular momentum plane in the t channel are a fixed

---

* See also E. L. FEINBERG's review at the 1964 Dubna conference.

pole, a moving pole and a moving branch point. It is now recognized that the theoretical objections raised against such singularities [18] are not valid for $t < 0$.

The latter model, which might be called centroperipheral, is undoubtedly more adequate than the multiperipheral one for the energy region up to 30 GeV. It allows for a variable proportion of central and peripheral collisions. Thus πp and $\bar{p}p$ would be mainly central, pp being mainly peripheral.

2. Shape of the diffraction peak in relation to inelastic processes. As mentioned above, all diffraction peaks measured up to now have in the interval

$$0 < -t \lesssim 1 \,(\mathrm{GeV}/c)^2 \quad \text{or} \quad 0 < -at \lesssim 10 \tag{II.1}$$

the same shape given by (I.14—15). This shape is best discussed in terms of the partial wave amplitude

$$\eta_l = 1 - e^{2i\delta_l} \tag{II.2}$$

where δ_l is the complex phase shift. The function η_l is real when the diffraction picture holds (purely imaginary scattering amplitude T). We first concentrate our discussion on this case. The effect of a non-vanishing ratio $\mathrm{Re}\,T/\mathrm{Im}\,T$ will be commented upon later.

The most conventional assumption concerning η_l, which is given by

$$\begin{cases} \eta = \eta_0 & \text{for} \quad 0 \le l \le L \\ \eta_l = 0 & \text{for} \quad l > L \end{cases} \tag{II.3}$$

and corresponds to absorption by a black sphere, is in clear disagreement with the facts. It gives

$$\frac{b}{a^2} \cong -0.08 \tag{II.4}$$

and produces a diffraction minimum for $-at \cong 3.7$. This disagreement is not suppressed by a slight rounding off of the edge at $l = L$.

Much better agreement is obtained with a Gaussian distribution

$$\eta_l = \eta_0 \exp(-l^2/L^2) \tag{II.5}$$

which gives a vanishing b. The ratio $\sigma_{\mathrm{el}}/\sigma_T$ (σ_{el} = total elastic cross-section) ranges from 0 to 0.25 as η_0 grows from 0 to 1[*]. The experimental values of $\sigma_{\mathrm{el}}/\sigma_T$ being of order 0.2 (see below), one concludes that η_0 is a little smaller than 1.

A further improvement is obtained if η_l is taken to be the smaller root of the equation

$$\eta_l - \frac{1}{2}\eta_l^2 = f_0 \exp(-l^2/L^2) . \tag{II.6}$$

This gives a positive b/a^2 ranging from 0 to 0.02 as f_0 grows from 0 to the maximum value $\frac{1}{2}$ allowed by unitarity. As to the ratio $\sigma_{\mathrm{el}}/\sigma_T$, it increases from 0 to a maximum of 0.185. Equation (II.6) is a special case of

---

[*] In contrast herewith, the black sphere distribution (II.3) gives a ratio $\sigma_{\mathrm{el}}/\sigma_T$ ranging from 0 at 0.5 as η_0 increases from 0 to 1.

the unitarity condition for elastic scattering which can be put in the form

$$\operatorname{Re}\eta_l - \frac{1}{2}|\eta_l|^2 = f_l \tag{II.7}$$

The left-hand side is the contribution of the elastic channel to the unitarity condition, the right-hand side being the total contribution of all inelastic channels. The precise definition of f_l is given by the equation

$$\langle \psi(k'_A k'_B) \mid \psi(k_A k_B) \rangle = \frac{\delta(P' - P)\sqrt{s}}{2\pi k \, \varepsilon_A \, \varepsilon_B} \sum_{l=0}^{\infty} (2l + 1) \, f_l \, P_l(\cos\theta) \tag{II.8}$$

$\psi(k_A k_B)$ = inelastic final state produced by initial state $|k_A k_B\rangle$ where k_A, k_B are the momenta of the incident particles A, B[*].

k = length of momentum of incident particles in c.m. system.

$\varepsilon_A, \varepsilon_B$ = energies of incident particles in c.m. system.

θ = angle between k_A and k'_A in c.m. system.

$\delta(P' - P)$ = four-dimensional delta function for total momentum-energy conservation.

The Gaussian l dependence adopted for f_l in (II.6) can be shown to hold true when the inelastic final state is mainly composed of many-particle states with small transverse momenta and with weak correlations between the particles. This model therefore proposes an explanation of the diffraction peak shape in terms of general properties of the dominant inelastic collisions[**], [***].

As shown by COTTINGHAM and PEIERLS [22], the model just discussed gives an excellent description of πp diffraction scattering at $p_L \gtrsim 15\,\text{GeV}/c$ where the real part of the scattering amplitude is sufficiently small to neglect its contribution to (II.7)[†]. The parameter f_0 is fixed by the ratio σ_{el}/σ_T which is 0.16, somewhat below the allowed maximum 0.185.

For pp collisions the ratio σ_{el}/σ_T is larger than 0.185 (its value is 0.276 ± 0.008 at $p_L = 12.8\,\text{GeV}/c$ and 0.244 ± 0.12 at $19.6\,\text{GeV}/c$). COTTINGHAM and PEIERLS attributed this to a non-vanishing real part of the scattering amplitude T, the existence of which was also predicted by dispersion relations [23] and was confirmed experimentally (see Fig. 3). By giving to this real part a short range they were able to account simultaneously for the observed pp elastic scattering at small and at large $|t|$ values [22].

Since we expect the diffraction picture to become valid at high energy, the applicability of the model (II.6) to pp scattering requires that the ratio σ_{el}/σ_T decrease below 0.185 at very high energy. It is interesting in this connection that, using a simultaneous extrapolation of pp and $\bar{\text{p}}$p data, CZYZEWSKI et al. find for the asymptotic limit of this ratio the value 0.178 ± 0.006 [13].

---

[*] The normalization is such that $\langle k'_A k'_B \mid k_A k_B \rangle = \delta(k'_A - k_A)\,\delta(k'_B - k_B)$.

[**] A preliminary discussion of this method and of the relation between inelastic collisions and diffraction peak width, was given in [20].

[***] A concrete, bremsstrahlung-type model of weakly correlated inelastic final states has recently been discussed for proton-proton collisions by BIAŁAS and RUIJGROK [21]

[†] This contribution is of order $(\operatorname{Re} T/\operatorname{Im} T)^2$, which is a few percent for very small $|t|$ and can be safely assumed to remain small for larger $|t|$.

III. Inelastic Two-body Collisions

Among the inelastic collisions, considerable interest is attracted by two-body collisions $A + B \to A' + B'$ where particles A' and/or B' are different from A and B. In fact A', B' can be unstable resonances of definite quantum numbers. The cross-section for such processes is surprisingly high at low and intermediate energies ($p_L \lesssim 5 \text{ GeV}/c$). For example in $K^+ p$ collisions at $p_L = 3 \text{ GeV}/c$, about 40% of the $K^0 \pi^+ p$ final states are in the two-body channel $K^{*+} p$ and 40% in the two-body channel $K^0 N^{*++}$ whereas only 20% are genuine three-body states [24]. It would be desirable to find a general, qualitative explanation for this remarkable fact. As the energy increases, it appears to be a general rule that the cross-section for any individual inelastic two-body process $A + B \to A' + B'$ decreases rapidly, at least when the internal quantum numbers of A' and/or B' differ from those of A and B. Typical examples are a cross-section decreasing like s^{-1} for $\pi^- p \to \pi^0 n$ and one decreasing like s^{-2} for $\pi p \to \rho p$. The number of two-body channels increases with s, however, and it is not known whether the total cross-section into all two-body channels decreases or becomes constant. Experimental indications on this point would be very helpful for our general understanding of high energy collisions. Thus, the success of the model discussed in Section I.2 and based on Eq. (II.6) is readily understood if at high energy most inelastic collisions are of many-body type. Whether the Gaussian l dependence of f_l can also be derived on general grounds when an appreciable fraction of the inelastic collisions leads to two-body states is an unsolved question.

1. One meson exchange models for inelastic two-body collisions. The model most commonly used to describe inelastic two-body collisions is the one meson exchange model (OME) with various improvements, mainly form factors [25] and absorption corrections [26]. Very extensive work has been devoted to these models. We shall limit ourselves to a few brief comments because excellent reviews have been given recently by Jackson [27] and Jackson et al. [28].

For processes dominated by one pion exchange the OME model, either with form factors or with absorption, has reached an appreciable degree of success. A favourable example is provided by the reaction $\pi p \to \rho p$ where both models give a very successful description of the differential cross-section as function of p_L and t in the intervals $2 < p_L < 8 \text{ GeV}/c$, $0 < -t < 0.5$ $(\text{GeV}/c)^2$. The absorption model has the advantage of predicting decay correlations and a non-vanishing difference between $\pi^+ p \to \rho^+ p$ and $\pi^- p \to \rho^- p$, quantities which are absent in the OME with or without form factors but are present in the data. The success of the one pion exchange models for nucleon-nucleon and antinucleon-nucleon collisions with isobar excitation is more limited and seems to be restricted to smaller energies [29].

More fundamental difficulties are encountered by the OME models in those cases, like $K p \to K^* p$, where vector-meson exchange is found to be required by the experimental data. Irrespective of whether form factors or absorption corrections are introduced, the model gives a cross-section independent of energy for $s \to +\infty$, in clear contradiction with the facts. The shape of the angular distribution, on the other hand, can be accounted for with more success, especially when absorption and form factor effects are included simultaneously. An example is given by charge exchange scattering, as discussed below.

A very interesting modification of the OME model, introducing the OME Born amplitudes in the K matrix rather than in the S matrix, has been recently studied by DIETZ and PILKUHN [30]. Although the method is quite different, an analysis of K^+p reactions shows that the main conclusions of good agreement with intermediate energy data and of difficulties with the energy dependence at large energy remain essentially valid. The model also allows to calculate elastic scattering and finds a satisfactory t dependence but too large an absolute value of the differential cross-section.

2. Charge exchange scattering. Among inelastic two-body collisions, charge exchange scattering processes like $\pi^-p \to \pi^0n$ and $K^-p \to K^0n$ are of particular interest because of their close relation to elastic scattering and also because accurate experimental information is becoming available. We shall discuss only meson-nucleon charge exchange which is

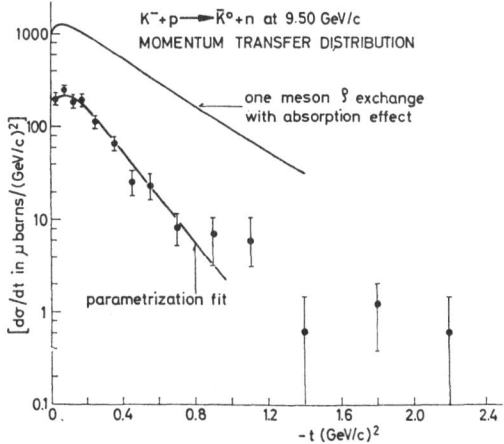

Fig. 6. $K^-p \to \bar{K}^0n$ differential cross-section at $p_L = 9.50$ GeV/c, following Ref. [31]

appreciably simpler than the proton-neutron case. Both K^-p and π^-p charge exchange data are available. They are characterized by the same general shape, as shown on Fig. 6 [31] and Fig. 7 [8]. This shape shows a remarkable difference from elastic scattering, in the form of a small maximum at very low momentum transfer. The most natural explanation is in terms of a spinflip amplitude of the same order of magnitude as the non-spin-flip amplitude (both being very small compared to the elastic amplitude which is believed to be mainly of non-spin-flip type). Beyond the small maximum the cross-section drops exponentially in $|t|$, with an exponent which is about the same as the exponent a in the elastic cross-section (I.14). At still larger $|t|$ there is indication of a second maximum decreasing much faster with energy than the first one.

Figure 6 illustrates, following H. Högaasen and J. Högaasen [32], how vector meson exchange corrected for absorption can reproduce the

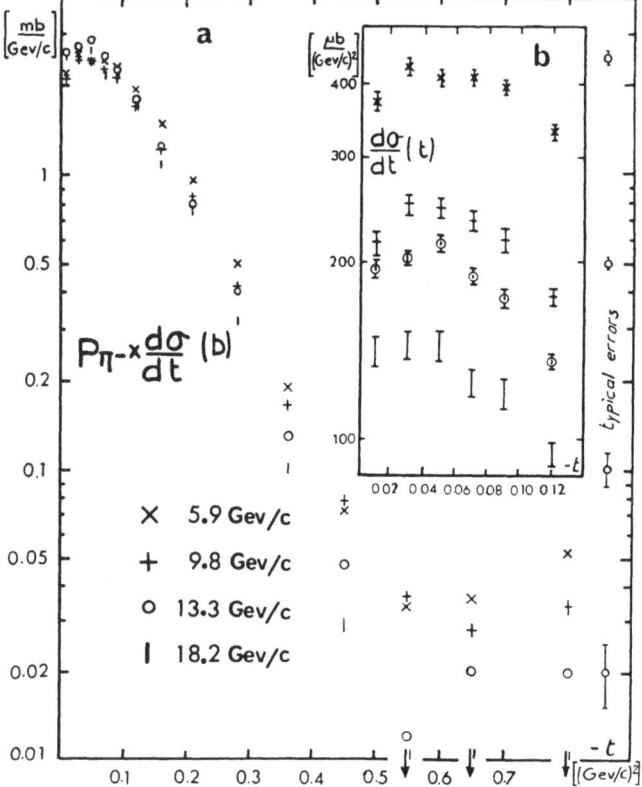

Fig. 7a and b. $\pi^-p \to \pi^0 n$ differential cross-section at four incident energies, following Stirling et al., Ref. [8]

general shape of the charge exchange data but neither the width nor the absolute magnitude. While these discrepancies can be reduced at moderate energies by the introduction of form factors, the constancy of the calculated cross-section with energy is certainly in strong disagreement with the facts.

The energy dependence of the charge exchange cross-section is known for the π^-p case [7, 8] and is illustrated in Fig. 7 where the ordinate is

$$p_L \frac{d\sigma}{dt} \sim s \frac{d\sigma}{dt} .$$

It raises the very interesting question of whether it is better interpreted by means of one single term or of two terms in an asymptotic expansion (I.6) for the charge exchange amplitude. Let us use first on single term

$$T_{\text{ch.ex.}} \cong c_1(t) \, s_t^{\alpha_1(t)} \, . \tag{III.1}$$

The data give $\alpha_1(0) \cong 0.5$ with good precision and suggest roughly $\alpha_1(-0.5) \cong 0$, $\alpha_1(-1) \cong -0.5$, t being measured in $(\text{GeV}/c)^2$. If the decrease of $\alpha_1(t)$ for $-t$ going from 0 to 1 is gradual, the single-term analysis (III.1) is quite satisfactory and the exponent $\alpha_1(t)$ has a large positive slope, of order $\alpha_1'(0) \cong 1$. If on the other hand, as the data suggest, $\alpha_1(t)$ changes slowly for $0 < -t < 0.5$, then drops rapidly for $-t$ around 0.5 to vary again more slowly for $-t > 0.5$, it would be more natural to use instead of (III.1) a two term expansion

$$T_{\text{ch.ex.}} \cong c_1(t) \, s_t^{\alpha_1(t)} + c_2(t) \, s_t^{\alpha_2(t)}. \tag{III.2}$$

with $\alpha_1(t)$ and $\alpha_2(t)$ changing slowly with t and having orders $\alpha_1 \cong 0.5$ and $\alpha_2 \cong -0.5$ for the whole t range. $c_1(t)$ should be more peaked than $c_2(t)$; then, at any energy s, there would be a value $t_0(s)$ such that the first term of (III.2) would dominate for $t_0 < t < 0$ and the second for $t < t_0$.

A decision between the two possibilities just discussed seems to be difficult with the statistics now available. It will probably have to await further experiments[**]. One point is already clear, however, namely that $\alpha_1(0) \cong 0.5$ in all cases. This explains the failure of the vector meson exchange model, which predicts $\alpha_1(0) = 1$, and is a clear and unambiguous success for the Regge pole model. Indeed, among all trajectories commonly considered in the Regge pole model, only the ρ trajectory has the quantum numbers involved in $\pi^- p$ charge exchange, and the model predicts a value of $\alpha_\rho(0)$ which is close to 0.5.

Isospin invariance implies that the crossed amplitude $\bar{T}_{\text{ch.ex.}}$ of the $\pi^- p$ charge exchange amplitude $T_{\text{ch.ex.}}$ verifies

$$\bar{T}_{\text{ch.ex.}}(st) = -T_{\text{ch.ex.}}(st) \, .$$

Comparing (I.6) and (I.7) we can then derive the phases of the coefficients $c_j(t)$ modulo π

$$c_j(t) = \pm |c_j(t)| \exp\left[-\frac{i\pi}{2} \left(\alpha_j(t) + 1\right)\right] \, . \tag{III.3}$$

This has interesting consequences for the polarization of the recoil neutron[*]. Assume (III.1) to hold with good accuracy both for the spin-flip and the non-spin-flip amplitudes. From (III.3) these amplitudes will have the same phase and zero polarization will result. In the higher approximation (III.2) polarization will occur through interference between the two terms of the expansion and its magnitude will decrease as $s_t^{\alpha_2(t)-\alpha_1(t)}$. Among the two possibilities mentioned above to account for the energy dependence of $d\sigma/dt$, the second suggested $\alpha_1 \sim 0.5$, $\alpha_2 \sim -0.5$. If it turns out to be valid, the polarization will decrease roughly as s^{-1}.

---

[*] The author acknowledges useful discussions with Professor G. Höhler and a very enlightening correspondence with Professor G. F. Chew on this question.

[**] Note added by the editor: a new analysis including additional data led to $\alpha(t) = 0.57 \pm 0.02 + (0.87 \pm 0.05)t$ for $t > -1.7 \ (\text{GeV}/c)^2$.

IV. Large Momentum Transfers — Higher Symmetries

We end by a few comments on some fundamental problems which begin to be tackled experimentally and theoretically from various sides, and which may well turn out to become in the future the most important items in the study of high-energy dynamics.

1. Large angle scattering. As discussed above, elastic scattering at small momentum transfer $[0 < -t < 1 \ (GeV/c)^2]$ appears to have the same properties for all hadrons and is probably just shadow scattering,

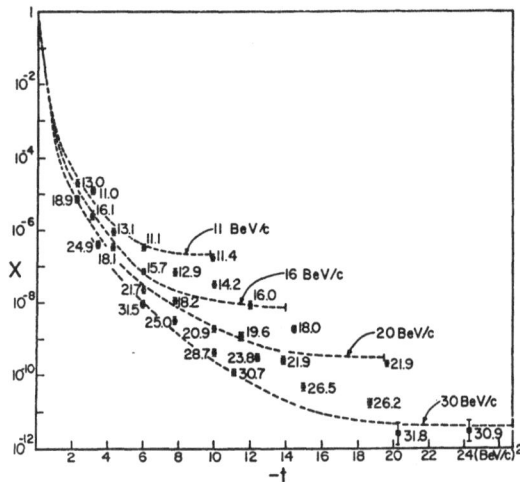

Fig. 8. pp large angle elastic scattering, following Ref. [33]

thereby reflecting only very little of the more specific features of high energy dynamics. One may hope that at large momentum transfers elastic scattering becomes more sensitive to the underlying dynamics. We are far from knowing whether this will be so, but a hopeful indication is given by the fact that pp and πp elastic scattering become different from each other at large angle. This is seen from Fig. 8, which gives the pp data at all energies available [33], and Fig. 9 giving the πp results at $p_L = 8 \ GeV/c$ [34]. Only one feature appears at present to be common to these results, namely the exponential drop of the differential cross-section at $\theta_{c.m.} \sim 90°$ with energy, roughly proportional to $\exp(-3.2\sqrt{s})$ * in agreement with the predictions of the statistical model [35]. It will be extremely interesting to study experimentally further cases of two-body elastic and inelastic processes at large momentum transfer. Only in this way will one be able to disentangle the many lines of interpretation which are possible at present. In addition to the statistical model, one has mentioned as possibilities a minimal interaction allowed by analyticity

---

* s is in $(GeV/c)^2$.

[36] and universal form factors [37]*, and there is little doubt that still other concepts could be introduced to analyze the same phenomena. Only experiment can decide about their usefulness. We note also that, as shown by COTTINGHAM and PEIERLS [22], the observed large angle pp scattering can be explained by giving a short range to the real

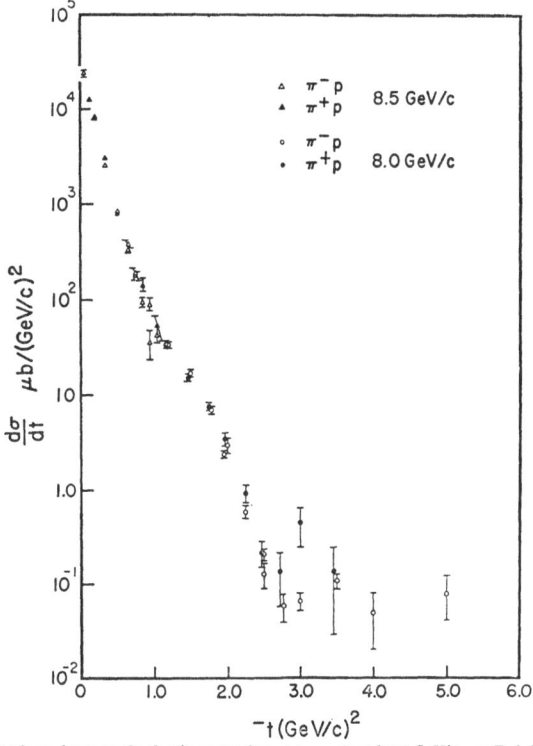

Fig. 9. πp large angle elastic scattering at $p_L = 8$ and 8.5 GeV/c, see Ref. [34]

part of the scattering amplitude, an explanation which does not apply to πp and may account for the very different shapes of the angular distributions.

Figure 10 shows πp large angle scattering plotted in $\cos\theta_{c.m.}$, up to the backward region which is reproduced in enlarged scale [34]. It reveals a pronounced backward peak for the π^+p system, with a surprisingly narrow width of the order of half the width of the forward diffraction peak. It is natural to associate this peak with nucleon exchange, and it is clear that a reggeized nucleon exchange model will have enough free parameters and functions to fit the data. It would be nice, however, to correlate this phenomenon with others in order to

---

* A modified and somewhat more intuitive picture of the Wu-Yang form factors can be obtained if one thinks of the baryons as bound states of three quarks and of the mesons as quark-antiquark bound states.

arrive at non-trivial conclusions. It should be recalled in this connection that the crossed reaction is $\bar{p}p \to \pi\pi$ [38]* so that a common analysis with help of expansions of type (I.6—7) can be carried out.

2. Unitary symmetry. $SU(3)$ and $SU(6)$ symmetry has been very successful in classifying particles and in predicting properties of three-particle vertex functions like those involved in electromagnetic and

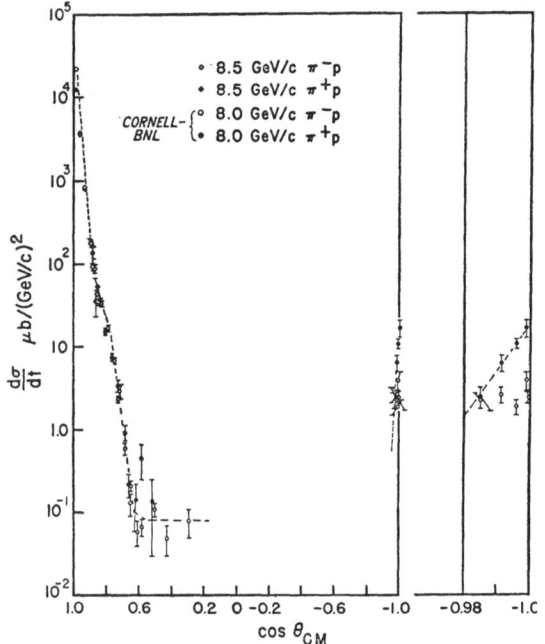

Fig. 10. πp large angle elastic scattering at $p_L = 8$ and 8.5 GeV/c, with special reference to backward region see Ref. [34]

weak interactions. The success is more limited for scattering and particle production processes, where it appears empirically that symmetry breaking effects considerably obscure the situation. A detailed discussion of this question has recently been given by HARARI [39]. At high energies, most elastic and inelastic collisions still involve small momentum transfers, so that the effect of symmetry breaking has no reason to become less important. Following Harari, we illustrate this fact by the following considerations.

Using U spin conservation one can demonstrate [40] that exact $SU(3)$ symmetry implies

$$|T(K^-p \to K^-p) - T(\pi^-p \to \pi^-p)| = |T(K^-p \to \pi^-\Sigma^+)| \quad \text{(IV.1)}$$

where all scattering amplitudes are taken at the same s and t. For $s \to +\infty$ at fixed t, the ratio

$$T(K^-p \to \pi^-\Sigma^+)/T(K^-p \to K^-p) \quad \text{(IV.2)}$$

---

* In equation (2) of Ref. [38] a factor 2 is missing in the right hand side (private communication from Dr. A. BIAŁAS).

tends rapidly to zero because it involves the strangeness exchange reaction $K^- p \to \pi^- \Sigma^+$. This implies

$$[T(K^- p \to K^- p)/T(\pi^- p \to \pi^- p)] \to 1 . \qquad (IV.3)$$

Taking this relation for $t = 0$ and applying the optical theorem, one concludes that it contradicts the fact that $K^- p$ and $\pi^- p$ have different total cross-sections at high energy. In other words, for s sufficiently large to neglect the ratio (IV.2), any difference between the left-hand side of (IV.3) and unity is due to symmetry breaking.

This has an important consequence for the relations proposed by JOHNSON and TREIMAN [41] to test $SU(6)$ symmetry

$$\frac{1}{2}[f(K^+ p) - f(K^- p)] = f(\pi^+ p) - f(\pi^- p) = f(K^0 p) - f(\overline{K}^0 p) \qquad (IV.4)$$

where f denotes the forward elastic scattering amplitude. As soon as the energy is high enough for strangeness exchange processes to become negligible, all quantities

$$\frac{f(K^- p)}{f(\pi^- p)}, \quad \frac{f(K^+ p)}{f(\pi^+ p)}, \quad \frac{f(K^+ p)}{f(K^- p)}, \quad \frac{f(K^0 p)}{f(\overline{K}^0 p)}$$

differ from 1 only through $SU(3)$ symmetry breaking effects [39]. Under these circumstances, any equality involving non-vanishing experimental values of the first and last differences in (IV.4) is a relation between $SU(3)$ violating terms and is no test of $SU(3)$ nor of $SU(6)$. Consequently, the Johnson-Treiman relation can only be used at low energy where strangeness exchange is appreciable and where one might hope that $SU(3)$ symmetric contributions to (IV.4) would no longer be overshadowed by the symmetry breaking effects. Whether this hope is justified is unknown.

We remain with the central question whether unitary symmetry will manifest itself more clearly in processes where not only energy but also all relevant momentum transfers are large compared to the baryon masses. Experimentally, one should for example check the triangular inequalities between cross-sections following from (IV.1), and many other similar tests of $SU(3)$, in two-body processes at high energy and large angle. Theoretically one should extract from future experiments a more definite picture of the dynamical behaviour of large momentum transfer phenomena, and guess how it accommodates $SU(3)$ symmetry. The undeniable progress made in the last years in the phenomenological study of high energy phenomena at low momentum transfers gives us good hope that similar advances will be possible in the more fundamental but less accessible region where all kinematical variables are large*.

---

* The success of $SU(3)$ and $SU(6)$ symmetry suggests that baryons might be bound states of three triplets, quarks for example, mesons being quark-antiquark bound states. This may give a simple explanation for the large proportion of inelastic two-body collisions at relatively high energies (see Section III), the reason being that in these collisions the incoming composite systems are simply excited, without creation of further quark anti-quark pairs. A discussion of the suppression of particle production in the quark model is given in [42].

References

[1] MANDELSTAM, S.: Nuovo cimento 30, 1127, 1148 (1963).
[2] GLASER, V.: Private communication.
[3] HOVE, L. VAN: Lectures at the Seminar on High Energy Physics and Elementary Particles, International Centre for Theoretical Physics of I.A.E.A., Trieste, May—June 1965, to be published
[4] LINDENBAUM, S. J.: Rapporteur's talk. International Conference on High Energy Physics, Dubna, August 1964
[5] FOLEY, K. J., R. S. GILMORE, R. S. JONES, S. J. LINDENBAUM, W. A. LOVE S. OZAKI, E. H. WILLEN, R. YAMADA, and L. C. L. YUAN: Phys. Rev. Letters 14, 862 (1965)
[6] BELLETTINI, G., G. COCCONI, A. N. DIDDENS, E. LILLETHUN, J. PAHL, J. P. SCANLON, J. WALTERS A., M. WETHERELL, and P. Zanella: Phys. Letters 14, 164 (1965)
[7] MANNELLI, I., A. BIGI, R. CARRARA, M. WAHLIG, and L. SODICKSON: Phys. Rev. Letters 14, 408 (1965)
[8] STIRLING, A. V., P. SONDEREGGER, J. KIRZ, P. FALK-VAIRANT, O. GUISAN, C. BRUNETON, P. BORGEAUD, M. YVERT, J. P. GUILLAND, C. CAVERZASIO and B. AMBLARD: Phys. Rev. Letters 14, 763 (1965)
[9] POMERANCHUK, I. IA.: Zhur. Eksp. Teor. Fiz. 34, 725 (1958); [English translation: Soviet Phys. — J.E.T.P. 7, 499 (1958)]
[10] CONZETT, H. E.: Phys. Letters 16, 189 (1965).
[11] BIAŁAS, A., and E. BIAŁAS: CERN preprint TH. 504 and Nuovo cimento 37, 1686 (1965)
[12] HOVE, L. VAN: Phys. Letters 5, 252 (1963); LOGUNOV, A. A., NGUYEN VAN HIEU, I. T. TODOROV, and O. A. KHRUSTALEV: Phys. Letters 7, 69 (1963)
[13] CZYZEWSKI, O., B. ESCOUBÈS, Y. GOLDSCHMIDT-CLERMONT, M. GUINEA-MOORHEAD, D. R. O. MORRISON, and S. DE UNAMUNO-ESCOUBÈS: Phys. Letters 15, 188 (1965)
[14] HOVE, L. VAN: Phys. Letters 7, 76 (1963)
[15] PHILLIPS, R. J. N., and W. RARITA: Lawrence Radiation Laboratory preprint UCRL 16033, Phys. Rev. 139, B 1336 (1965)
[16] AMATI, D., A. STANGHELLINI, and S. FUBINI: Nuovo cimento 26, 896 (1962)
[17] DREMIN, I. M., I. I. ROYSEN, E. L. FEINBERG, D. S. CHERNASVSKI, and R. WHITE: Proceedings, 1964 International Conference on High Energy Physics at Dubna, to be published
[18] GRIBOV, V. N.: Nuclear Phys. 22, 249 (1961); OEHME, R.: Phys. Rev. Letters 9, 358 (1962)
[19] HOVE, L. VAN: Nuovo cimento 28, 798 (1963) and Revs. Modern Phys. 36, 655 (1964).
[20] HOVE, L. VAN: Physikertagung Stuttgart 1962, p. 84. Mosbach/Baden: Physik-Verlag 1963
[21] BIAŁAS, A., and TH. RUIJGROK: CERN preprint TH. 535 and Nuovo Cimento, to be published
[22] COTTINGHAM, W. N., and R. F. PEIERLS: Phys. Rev. 137, B147 (1965)
[23] SÖDING, P.: Phys. Letters 8, 285 (1964)
[24] LYNCH, G. R., M. FERRO-LUZZI, R. GEORGE, Y. GOLDSCHMIDT-CLERMONT, V. P. HENRI, B. JONGEJANS, D. W. G. LEITH, F. MÜLLER, and J. M. PERREAU: Phys. Letters 9, 359 (1964) and Phys. Rev., to be published
[25] FERRARI, E., and F. SELLERI: Suppl. Nuovo cimento 24, 453 (1962)
[26] SOPKOVICH, N. J.: Nuovo cimento 26, 186 (1962)
[27] JACKSON, J. D.: Revs. Modern Phys. 37, 484 (1965)
[28] —, J. T. DONOHUE, K. GOTTFRIED, R. KEYSER, and B. E. Y. SVENSSON: Phys. Rev. 139, B 428 (1965)
[29] FERRARI, E., S. GENNARINI, and P. LARICCIA: CERN preprint TH. 521 (to be published in Nuovo cimento); SVENSSON, B. E. Y.: CERN preprint TH. 549 (to be published in Nuovo cimento)
[30] DIETZ, K., and H. PILKUHN: Nuovo cimento 37, 1561 (1965) and 39, 928 (1965)

[31] ASTBURY, P., G. FINOCCHIARO, A. MICHELINI, C. VERKERK, C. H. WEST, W. BEUSCH, B. GOBBI, M. PEPIN, M. A. POUCHON, and E. POLGAR: Phys. Letters 16, 328 (1965)
[32] HÖGAASEN, H., and J. HÖGAASEN: Nuovo cimento 39, 941 (1965)
[33] COCCONI, G., V. T. COCCONI, A. D. KRISCH, J. OREAR, R. RUBINSTEIN, D. B. SCARL, B. T. ULRICH, W. F. BAKER, E. W. JENKINS, and A. L. READ: Phys. Rev. 138, B165 (1965)
[34] OREAR, J., R. RUBINSTEIN, D. B. SCARL, D. H. WHITE, A. D. KRISCH, W. R. FRISKEN, A. L. READ, and H. RUDERMAN: Phys. Rev. Letters (1965)
[35] FAST, G., and R. HAGEDORN: Nuovo cimento 27, 208 (1963); FAST, G., R. HAGEDORN, and L. W. JONES: Nuovo cimento 27, 856 (1963); BIAŁAS, A., and V. F. WEISSKOPF: Nuovo cimento 35, 1211 (1965); HAGEDORN, R.: Nuovo cimento 35, 216 (1965)
[36] CERULUS, F., and A. MARTIN: Phys. Letters 8, 80 (1964); KINOSHITA, T.: Phys. Rev. Letters 12, 256 (1964); MARTIN, A.: Nuovo cimento 37, 671 (1965)
[37] WU, T. T., and C. N. YANG: Phys. Rev. 137, B708 (1965)
[38] BIAŁAS, A., and O. CZYZEWSKI: Phys. Letters 13, 337 (1964)
[39] HARARI, H.: Lectures at the Seminar on High Energy Physics and Elementary Particles. International Centre for Theoretical Physics of I.A.E.A., Trieste, May—June 1965, to be published
[40] —, and H. J. LIPKIN: Phys. Rev. Letters 13, 208 (1964)
[41] JOHNSON, K., and S. B. TREIMAN: Phys. Rev. Letters 14, 189 (1965)
[42] KOBA, Z.: Phys. Letters 16, 326 (1965)

Prof. Dr. L. VAN HOVE
Theory Division
CERN
Geneva 23 (Switzerland)

Recent High Energy Electron Investigations at Stanford University*

C. D. Buchanan, H. Collard,
C. Crannell, R. Frosch, T. A. Griffy,
R. Hofstadter, E. B. Hughes,**
G. K. Nöldeke, R. J. Oakes,
K. J. van Oostrum. R. E. Rand,
L. Suelzle and M. R. Yearian,
Stanford University, Stanford, California,
B. C. Clark and R. Herman,
General Motors Research Laboratories,
Warren, Michigan,
D. G. Ravenhall,
University of Illinois, Urbana, Illinois

(Presented by R. HOFSTADTER)

I. Introduction

The purpose of this paper is to bring to this Conference an up-to-date summary of recent developments in high energy electron physics at Stanford. The material can be divided into the following topics:
A. Nucleon Form Factors
B. Deuteron Form Factors
 1. Coincidence Studies of Elastic Scattering
 2. Inelastic Continuum Studies
C. Nuclear Physics
 1. Ca^{40}, Ca^{44} Investigations
 2. Study of the Inelastic Continuum in He^3
 3. Magnetic Octupole Moments in Light Nuclei
D. Shower Studies at 900 MeV.

Since the above material covers so many topics I shall have to be very brief in discussing each topic. Separate papers will be published subsequently on the individual subjects and these will have the appropriate authors' names on them. In the meantime, and for convenience, I have grouped all the authors names under a single title but it is obvious that not all individuals have been concerned with each part of the research.

* Work supported in part by the Office of Naval Research of the U.S. Navy (Contract Nonr 225 (67)), by the Office of Scientific Research of the U.S. Air Force, by the U.S. National Science Foundation, and also by a grant of the Research Laboratories, General Motors Corporation.
** Present address: Princeton University, Princeton, New Jersey.

II. Results

A. Nucleon Form Factors

Some of the material we shall show in the figures has been presented in preliminary form at the Dubna Conference but is thus far unpublished. We present herewith our final results on the proton and neutron form factors. This material will also appear shortly in the Physical Review [1].

The elastic cross sections have been obtained from curves such as shown in Fig. 1, which presents an elastic electron scattering peak at the

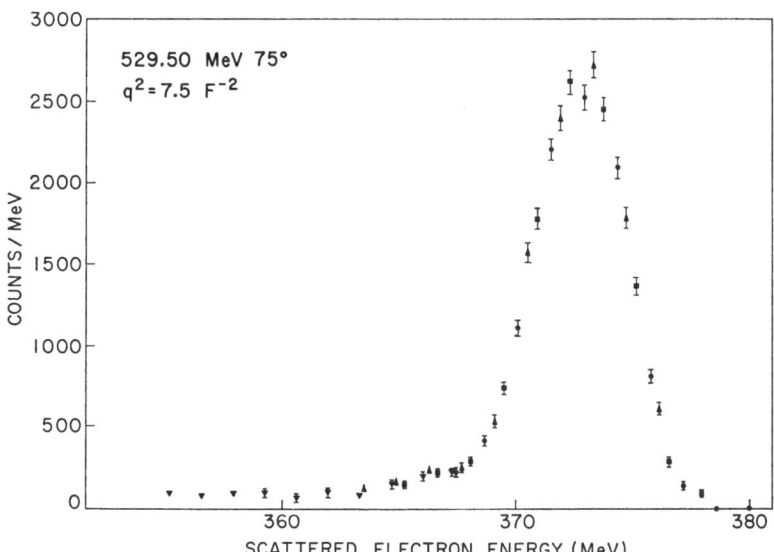

529.50 MeV 75°
$q^2 = 7.5$ F^{-2}

Fig. 1. A typical elastic electron-proton scattering peak

indicated energy and angle on a thin liquid hydrogen target. Fig. 2 shows comparable data for the deuteron, as well as a comparison hydrogen peak. From such data conventional analysis using final state corrections enable the four nucleon form factors, G_{Ep}, G_{Mp}, G_{Mn}, and G_{En} to be obtained. The results are all grouped together in Fig. 3 and show that within experimental error, the previously noted equalities hold

$$G_{Ep} \cong G_{Mp}/2.79 \cong G_{Mn}/(-1.91) \tag{1}$$

and also

$$G_{En} \cong 0. \tag{2}$$

These facts now seem to be finding a theoretical explanation for the first time in terms of symmetry considerations. A composite summary of form factors using the Stanford results as well as the work of other authors, is given in Figs. 4 and 5 [2—5]. Detailed numerical values as well as empirical three-pole CLEMENTEL-VILLI fitted expressions are available in an article by HUGHES et al. to be published in Physical Review [1]. The solid lines in Figs. 4 and 5 are the three pole fits made to only the present data. As may be seen, however, the fits satisfy other data as well. The accurate ω- and Φ-meson masses are used in the fit but

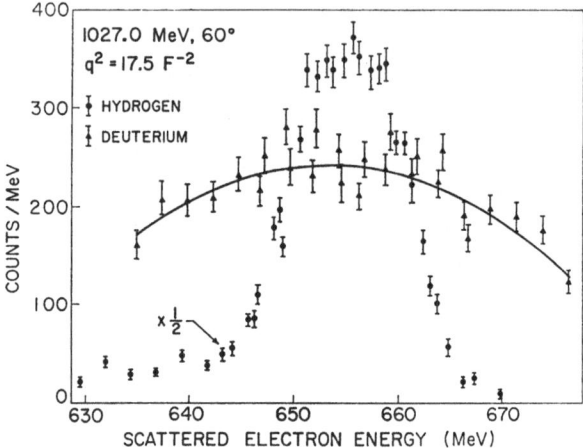

Fig. 2. Portion of a deuteron inelastic continuum near the peak, and a comparison proton peak

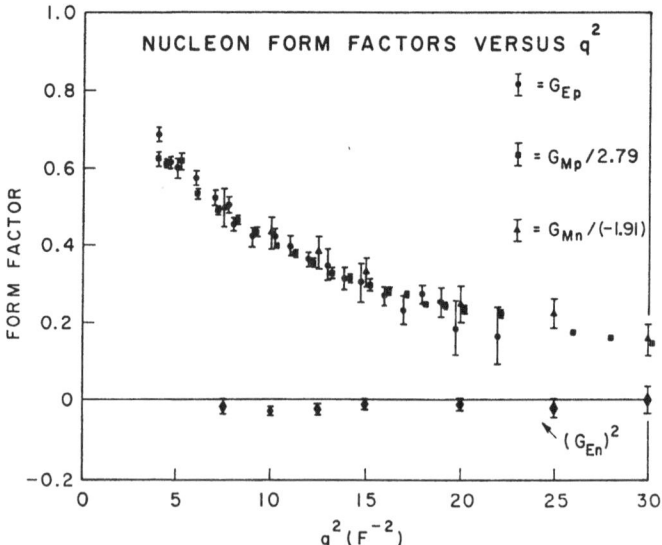

Fig. 3. The form factors for the proton and neutron. Note the approximate equality of three of the four form factors

the value of the ρ-meson mass empirically determined from the fit is 548 ± 24 MeV. The data are also consistent within experimental errors with the constraint that $G_{ES} = G_{MS}$ and $G_{EV} = G_{MV}$ at $q^2 = -4M^2$ where these G's are isotopic form factors. Also, with the exception of G_{MV}, the constant terms in the isotopic form factors are zero within the error of fitting.

It is possible to make a four-pole fit to the data using two isoscalar mesons (ω, Φ) and two isovector mesons $(\rho, ?)$. When this is done, using

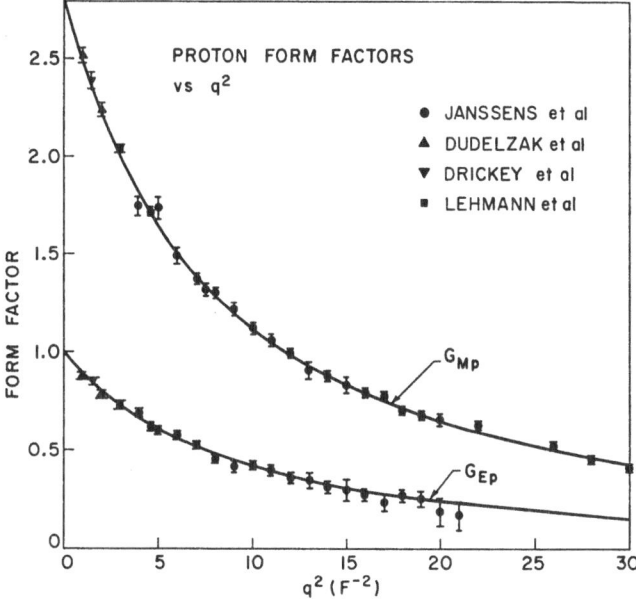

Fig. 4. Proton form factors from various authors. The solid lines refer to the three-pole fit described in the text

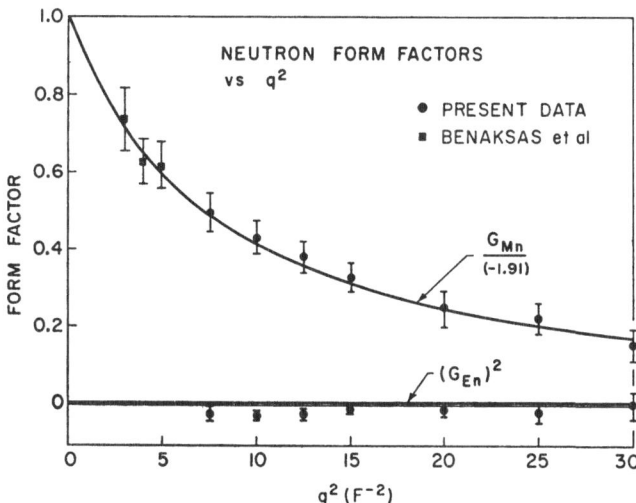

Fig. 5. Neutron form factors. The solid lines are from the three-pole fit described in the text

the exact *actual* values of the ω, Φ, ρ masses, a successful fit of the data furnishes a value of the unknown meson mass near 1200 MeV. Perhaps this is the B meson, but in any case, such a heavier meson is called for.

Except for the slight anomaly in obtaining negative $(G_{En})^2$ values, the nucleon form factor data appear to be in reasonably good shape. Improvements are needed, as is well known, in the theory of the deuteron *. It is anticipated that such improvements will lead to the explanation of the negative values of G_{En}^2.

B. Deuteron Form Factors

1. **Coincidence studies.** The previous remarks lead naturally to a further study of the deuteron. Only brief remarks will be made here on the coincidence method, since a more detailed report of this material will be

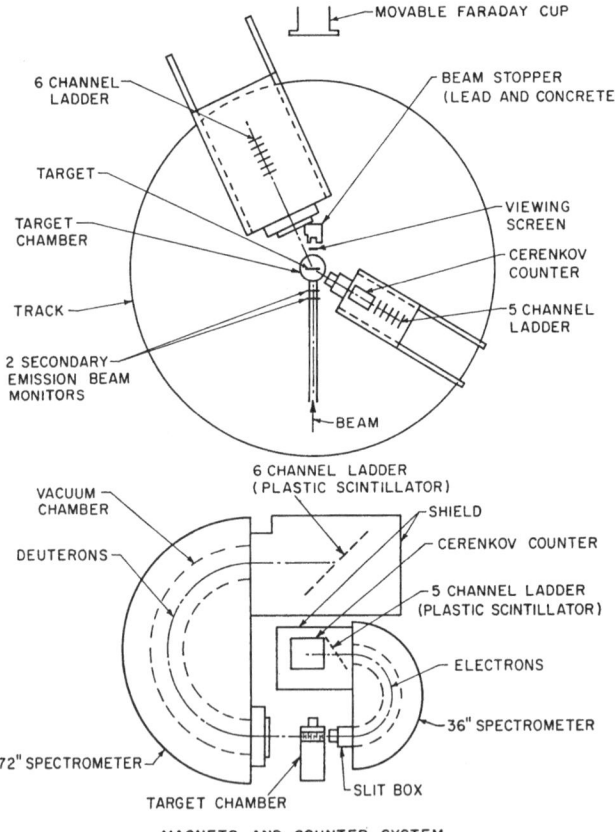

Fig. 6. The experimental arrangement used by Buchanan and Yearian to study electron-deuteron coincidences

* Note added at conference: M. Gourdin and his colleagues have presented to the conference an improved theory of the deuteron in which real values of G_{En} are obtained using older data of de Vries et al. [Phys Rev. **134**, B 848 (1946)]. (See the paper by M. Gourdin, M. Le Bellac, F. M. Renard and J. Trân Thanh Vân.)

given in a paper by C. D. BUCHANAN and M. R. YEARIAN [6] at this Conference.

The arrangement of spectrometers is shown in Fig. 6. Coincidences are required between the elements of two independent ladder counters, one ladder situated in each counter. Elastic scattering events can be selected by setting the spectrometers for the special conditions determined by the kinematics of elastic scattering.

The theory is represented by the following equations:

$$\frac{d\sigma}{d\Omega}\Big/\sigma_{NS} = A(q^2) + B(q^2)\tan^2\frac{\theta}{2} \tag{3}$$

$$A(q^2) = (1-\eta)^2[G_C^2 + G_Q^2] + \frac{8}{3}\eta\, G_M^2 \tag{4}$$

$$B(q^2) = \frac{16}{3}\eta\,(1+\eta)\, G_M^2 \tag{5}$$

where

$$\eta = q^2/4M_D^2 \tag{6}$$

and

$$G_C = G_{ES}I_C;\quad G_Q = G_{ES}I_Q,\quad G_M = G_{MS}I_{M1} + G_{ES}I_{M2} \tag{7}$$

in which G_{ES} and G_{MS} are the nucleon electric and magnetic isotopic scalar form factors and I_C, I_Q, I_{M1} and I_{M2} are integrals over the

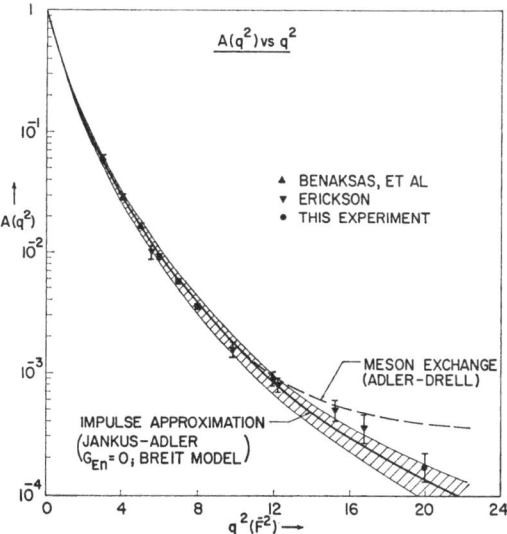

Fig. 7. Behavior of the coefficient A of the text in elastic scattering of electrons from the deuteron

deuteron's wave functions [7]. Other quantities have their conventional meanings.

The behavior of the quantities A and B is shown in Figs. 7 and 8. The figures also include earlier data of BENAKSAS et al. [8] and of E.

ERICKSON [7] obtained by detecting recoil deuterons in non-coincidence methods. A comparison of these results is made with the impulse approximation theory of JANKUS [9]. The uncertainty in obtaining the quantities A and B from theory is shown by the shaded area. It may be seen that the quantity A is in reasonable agreement with the Jankus theory, assuming the validity of $G_{En} = 0$ and the Breit set of deuteron parameters [7]. At

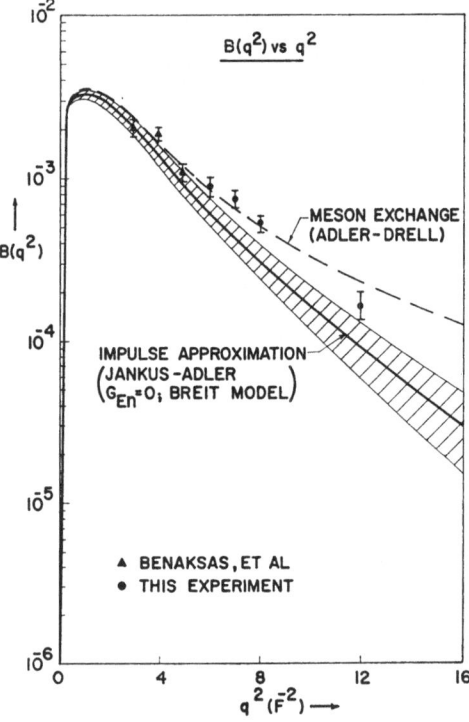

Fig. 8. Behavior of the coefficient B of the text in elastic scattering of electrons from the deuteron

larger values of q^2, for example $q^2 > 14$ F^{-2} there may even be a little evidence for departure of the data from theory. A theory of ADLER and DRELL [10] which assumes that a meson-exchange term contributes to the elastic-scattering predicts the dashed curves shown in Figs. 7 and 8. A diagram involving a constant $\rho - \pi - \gamma$ vertex makes the most important contribution to the correction. At $q^2 = 20$ F^{-2} the data appear to favor a Jankus theory not involving meson exchange, but since relativistic effects are taken into account only approximately in both theories, a decision between the two theories can probably not yet be made.

In the case of Fig. 8, where the behavior of B is shown, the data appear to favor the meson-exchange case at values of $q^2 > 6$ F^{-2} but the value of B at $q^2 = 12$ F^{-2} shows a tendency to be low. If one uses a form factor at the $\rho - \pi - \gamma$ vertex based on the mass of the ω meson the agreement is improved. However, further data are needed and a better estimate of relativistic effects still needs to be made.

2. Inelastic continuum studies. Let us first describe the work of HUGHES et al. [11] on the investigation of the area method in the deuteron to determine neutron form factors. This method is based on a

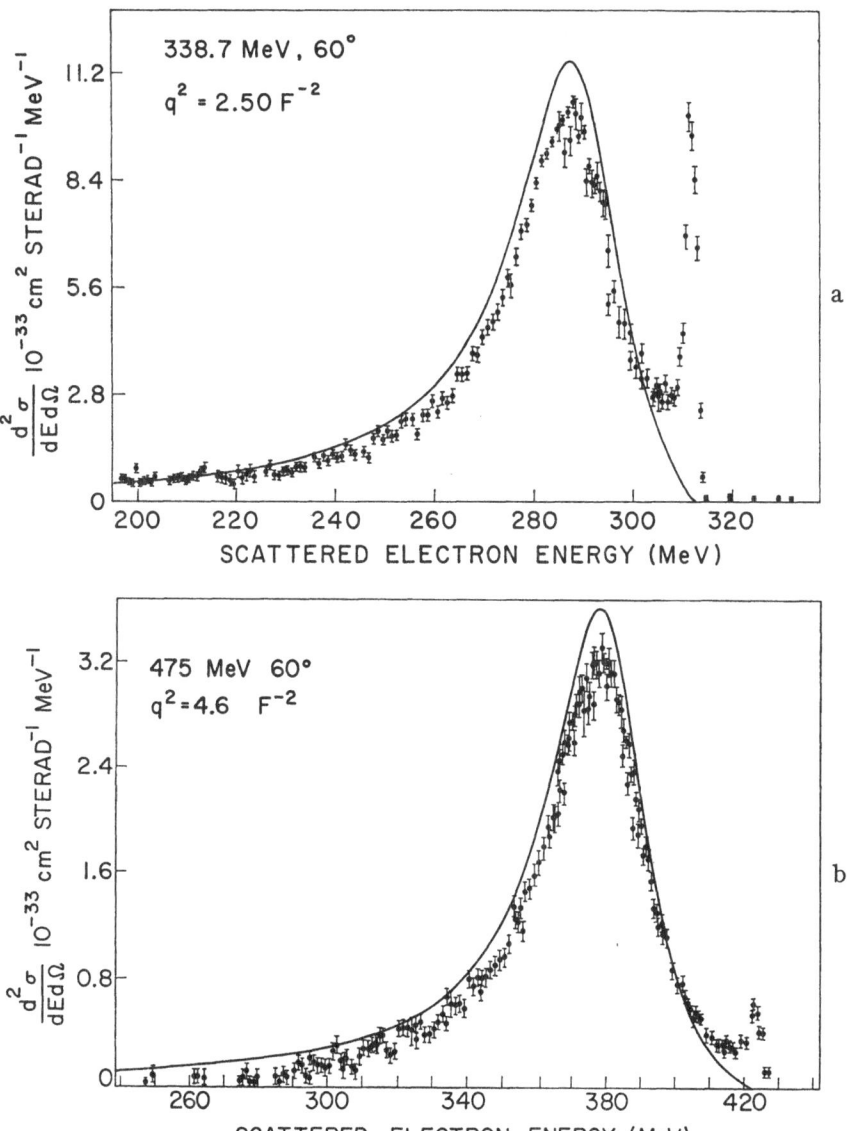

Fig. 9 a and b. Spectra of electrons scattered from the deuteron. Note the presence of the elastic peaks as well as the inelastic continua. The data were taken with the 72″ spectrometer and a 10-channel ladder

sum rule due to BLANKENBECLER [12]

$$\left(\frac{d\sigma}{d\Omega}\right)_{\text{area}} = (1 + \varDelta)\left\{\left(\frac{d\sigma}{d\Omega}\right)_{\text{p}} + \left(\frac{d\sigma}{d\Omega}\right)_{\text{n}}\right\} \tag{8}$$

where $\varDelta$ is a function whose numerical value is of the order of 0.02 or smaller. For these calculations, since $\varDelta$ is difficult to determine, it is omitted. Fig. 9 shows two examples of the data, along with simple deuteron curves calculated from Durand's theory [13], which assumes a Hulthén model for the $^3S_1$ part of the deuteron wave function. The deuteron wave function does not include D-state corrections, final state corrections or the effects of meson-exchange currents. The data show clearly that the peak method requires corrections in accord with expectations. The area method should be less sensitive than the peak method to final state etc. corrections and so this method can give information of an indirect kind to serve as a check on the various corrections

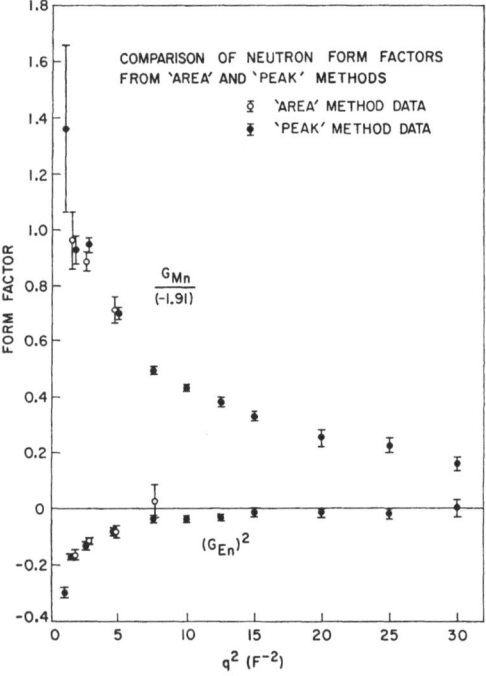

Fig. 10. Neutron form factors obtained from the "peak" and "area" methods. The open circles show the results of the area method

that influence the peak method. Fig. 10 shows some results of this method when applied to a determination of the neutron form factors. The open circles in the figure show the area method results and the other data give the peak method results. It may be seen that the results of the area method show the same tendencies as those of the peak method. It seems plausible therefore that the errors made in both methods are small. The detailed shape of the deuteron inelastic continua should be very valuable in determining a future correct deuteron theory.

C. Nuclear Physics

1. Investigations of Ca^{40} and Ca^{44}. The study of elastic scattering of electrons from Ca^{40} has been going on at Stanford for several years. The latest results by CROISSIAUX et al. [14] at 250 MeV give an accurate representation of the charge distribution in Ca^{40} in its ground state. The results are describable in terms of a Fermi distribution, which is a special case of Eq. (9)

$$\varrho_{c,z,n}(r) = \varrho \left\{ \exp \left(\frac{r^n - c^n}{z^n} \right) + 1 \right\}^{-1} \tag{9}$$

for which $n = 1$. The quantity c for the Fermi distribution is the radial distance to the 50% point and z is related to the conventional $(90-10\%)$ skin thickness, t, by the approximate formula

$$t = 4.40 \, z \, . \tag{10}$$

Eq. (9) also includes a Modified Gaussian distribution for which $n = 2$. For the Fermi distribution which gives an extremely good fit to the data, $c = 3.602$ F and $t = 2.51$ F. The Modified Gaussian gives a poorer fit for which the best values are $c = 3.73$ F and $t = 2.80$ F. When plotted both charge distributions look quite similar to each other. The quantities c and t are determined to within an accuracy of $\pm 1\%$ and $\pm 5\%$ respectively. For the Fermi model the accuracy may be somewhat higher.

One of the fascinating questions about nuclear matter is the problem of how a nuclear charge distribution changes upon the addition of uncharged neutrons to an already existing core. In other words how do different isotopes compare with each other? One of the best examples to study is calcium. For here are available several nuclei among which the most important are Ca^{40}, Ca^{44} and Ca^{48}. Ca^{40} and Ca^{48} are especially interesting since both are doubly magic and hence are spherical nuclei. Ca^{44} is magic in protons and since it lies between two other doubly magic nuclei, it is probably spherical or very nearly so. In going from Ca^{40} to Ca^{44} we add four neutrons without affecting the nucleus in any other way. In going from Ca^{40} to Ca^{48} we add 8 neutrons (out of 20) without affecting the nucleus in any other way. It is apparent that any theory of nuclear matter with a finite boundary will be subjected to a very searching test when confronted with the results of the experimental findings in the cases of the nuclei Ca^{40}, Ca^{44} and Ca^{48}. For example, do c and t both change, or does neither change, or does one change and the other not change? Also by how much do these quantities change? The usual rules for determining nuclear size and shape [15] tell us that t remains constant and c varies as $A^{\frac{1}{3}}$, viz.,

$$c = 1.06 \, A^{\frac{1}{3}} \, \text{F} \tag{11}$$

in the neighborhood of calcium. In going from Ca^{40} to Ca^{44} one might expect a change of 3.2% in c if rule (11) holds for isotopes. The half-maximum density point in the general charge distribution (9) is not

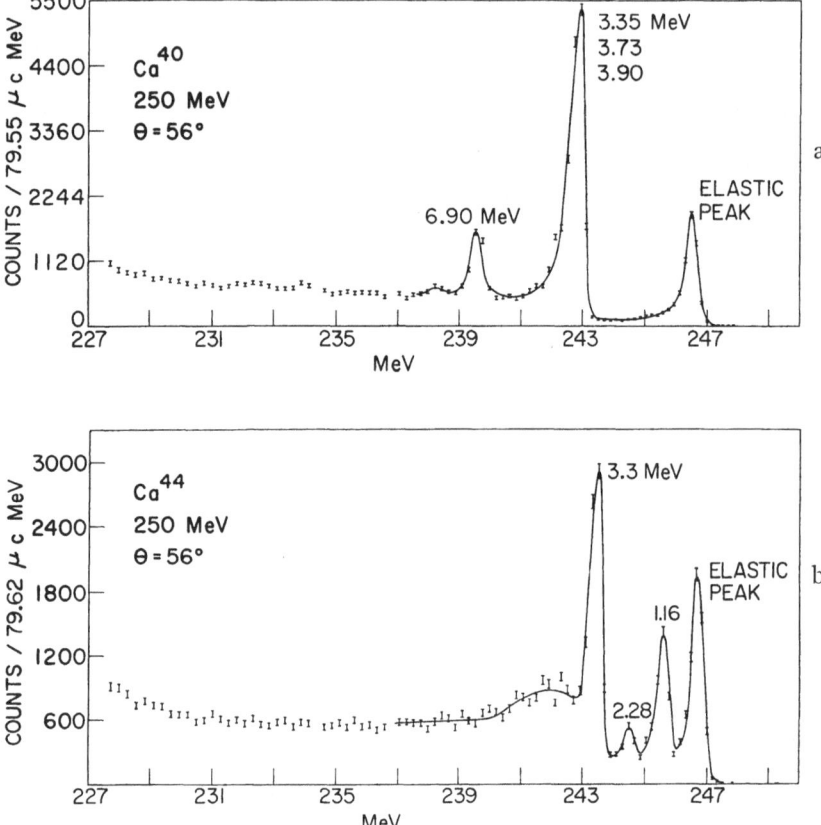

Fig. 11 a and b. Spectra of electrons scattered at 250 MeV from Ca^{40} and Ca^{44}. Note the scattering peaks from the excited states of these nuclei

exactly equal to c. We shall call the half-maximum density point $r_{0.5}$ in general.

Experiments have been carried out at 250 MeV simultaneously on two nearly identical metallic targets of Ca^{40} and Ca^{44} (B. C. CLARK et al. [17]). The preliminary results for Ca^{40} and Ca^{44} are shown in Figs. 11, 13, 14 and 15. Fig. 11 shows the energy spectra of scattered electrons under typical conditions. Note that the 1.16 MeV level in Ca^{44} is easily resolved. The data were obtained in the Stanford 72″ spectrometer with the 100 channel ladder detector of L. R. SUELZLE and M. R. YEARIAN [16]. This detector is shown in Fig. 12. In Figs. 14 and 15 we see the

quantity

$$\frac{\left(\dfrac{\mathrm{d}\sigma}{\mathrm{d}\Omega}\right)_{40} - \left(\dfrac{\mathrm{d}\sigma}{\mathrm{d}\Omega}\right)_{44}}{\left(\dfrac{\mathrm{d}\sigma}{\mathrm{d}\Omega}\right)_{40} + \left(\dfrac{\mathrm{d}\sigma}{\mathrm{d}\Omega}\right)_{44}} \tag{12}$$

plotted as a function of the scattering angle θ.

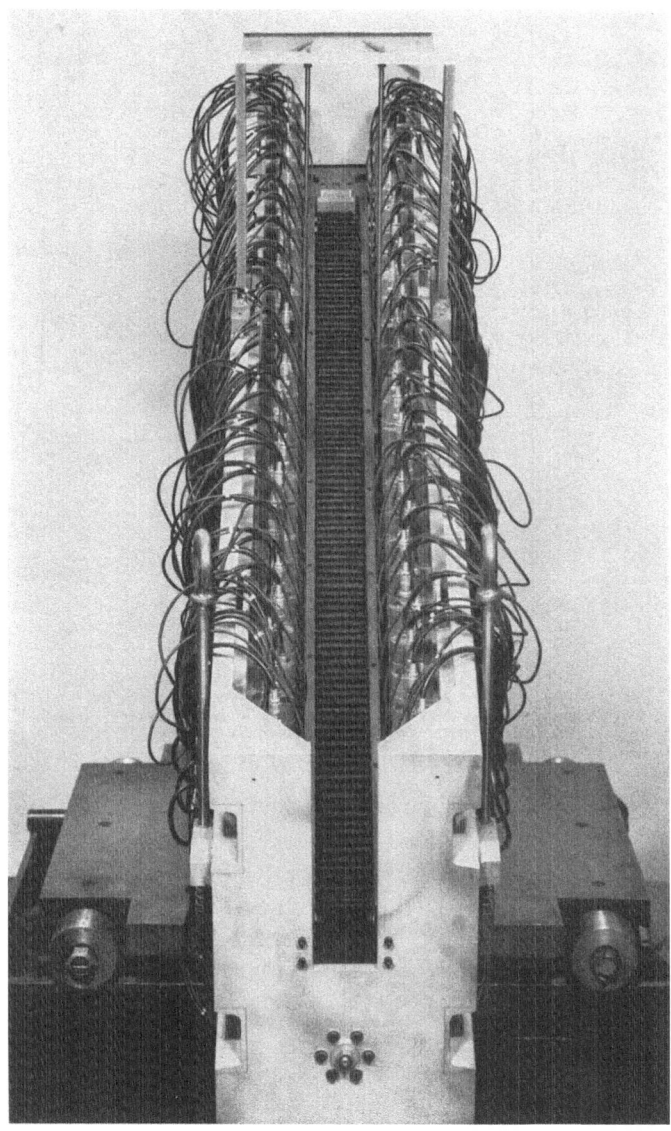

Fig. 12. A photograph of the 100 channel ladder used in the 72″ spectrometer

These data have been interpreted by B. C. CLARK, R. HERMAN and D. G. RAVENHALL [17] using a well established phase shift method. The solid and dashed lines in Figs. 14 and 15 show a comparison between the theoretical curves and the experimental data. From least square

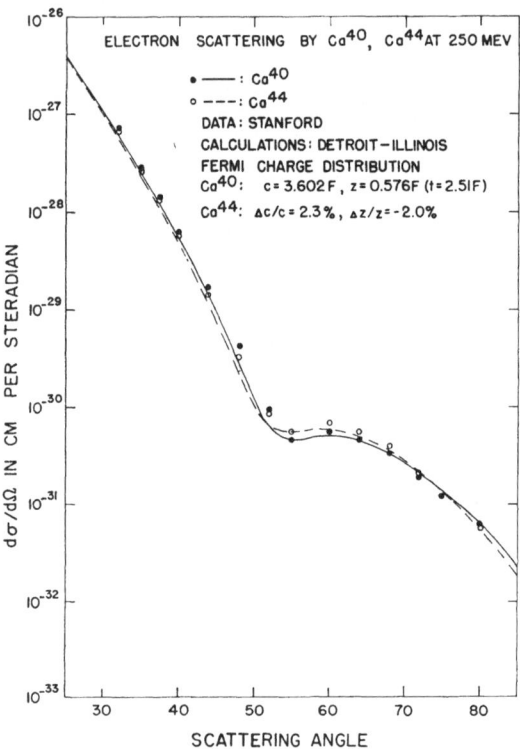

Fig. 13. Angular distributions for Ca⁴⁰ and Ca⁴⁴. The continuous curves show the results of recent theoretical calculations by CLARK et al. [17] for the Fermi model with the parameters shown. The absolute values are a little bit low on the average. See further comments on the absolute values in the text

fitting of the data one can conclude that for the Fermi model (Fig. 14) $r_{0.5}$ changes by $(+2.1 \pm 0.3)\%$ and t changes by $(-1.9 \pm 1.5)\%$ in going from Ca⁴⁰ to Ca⁴⁴. The corresponding results are $\dfrac{\Delta r_{0.5}}{r_{0.5}} = (+2.61 \pm 0.3)\%$ and $\dfrac{\Delta t}{t} = (-1.9 \pm 1.5)\%$ for a Modified Gaussian model. The Fermi results agree beautifully with what can be determined from the mu-mesic atom results shown in Fig. 14. Thus the change in c is not as large as would be predicted by the $A^{\frac{1}{3}}$ rule* and t is probably changing by a small

---

* Evidence for changes of effective nuclear radii smaller than predicted by the usual $A^{\frac{1}{3}}$ rule has been observed previously in both isotope shift and mu-mesic atom studies.

amount. It is very satisfying that the two different models agree so well in giving the same incremental changes.

In Fig. 13 we can see that the absolute measure of the cross section is still not exactly correct. This is a problem we have faced many times

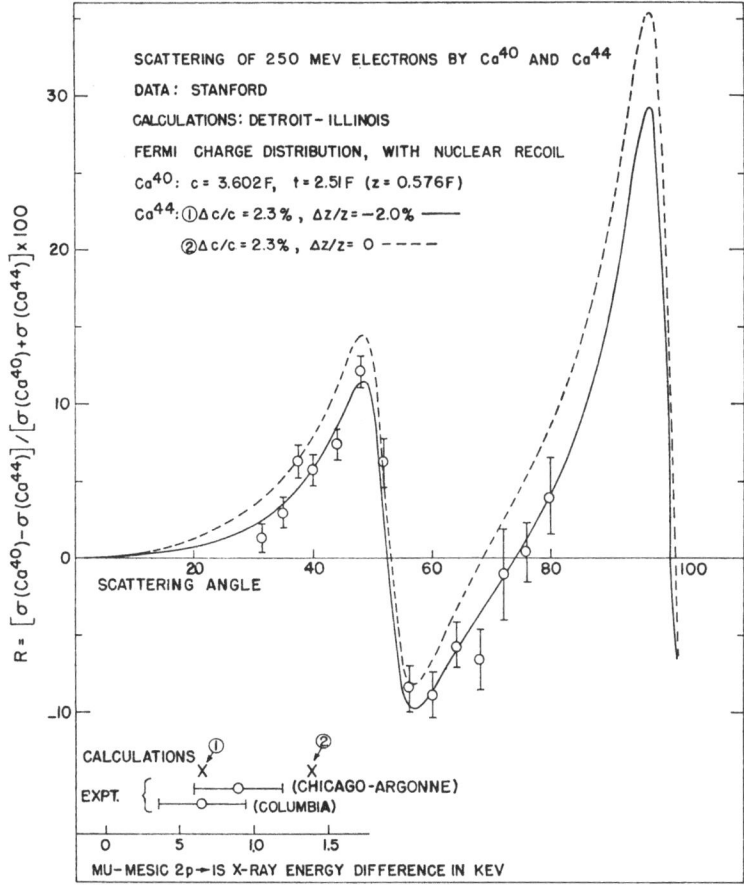

Fig. 14. The cross section difference, $(Ca^{40}—Ca^{44})$, divided by the sum, $(Ca^{40} + Ca^{44})$, is plotted as ordinate. These ratios are independent of absolute values. The Fermi model is compared with experiment. Mu-mesic information is given at the bottom of the figure

previously and these new results must be weighed together with older results until our technique of absolute measurement is improved. In the meantime we wish to emphasize that the relative data of Figs. 14 and 15 are nearly independent of the absolute values of the cross section. The theoretical interpretation is equally independent of the absolute values, so that the incremental changes noted above are likely to stand until their accuracy can be improved further. As we pointed out previously the incremental results must be explained by any correct theory of nuclear matter.

Experiments are being prepared for a study of Ca^{48} and a target of this material is now being fabricated by the Oak Ridge Target Fabrication Group. It is interesting to speculate on what will happen in the case of Ca^{48}.

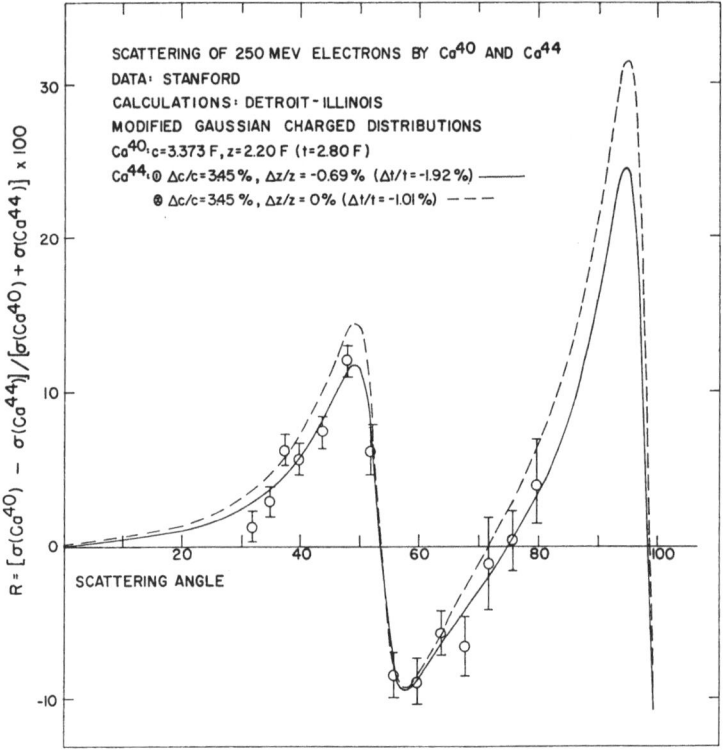

Fig. 15. This figure is similar to Fig. 14 but refers to a Modified-Gaussian model

2. Inelastic continuum in He³. Experiments have been carried out in which electrons are scattered from high pressure targets of He³. The inelastic spectra have the appearance shown in Fig. 16. More recent spectra show much more detail but are not yet ready for presentation. Positive particle counting rates were also measured in order to estimate the number of π^- mesons under the low energy tail of the inelastic peak. Assuming a ratio of π^-/π^+ to be approximately the same as in deuterium (except for a factor of 2 to allow for the two protons in He³) it is possible to correct for the number of π^- mesons contaminating the scattered electron continuum. In all cases the data are considered reliable only when the pion contamination is not too large. The number of electrons scattered after an $e - \pi$ process are also estimated and subtracted. In this way Collard [18] obtains the corrected curves labeled "experiment" in Fig. 17. Also shown in Fig. 17 are theoretical curves calculated by Oakes and Griffy [19] as described below.

These authors have calculated the spectrum of inelastic electrons scattered in the process of electrodisintegration of He³ by assuming that the electrons interact with a single "quasi-free" nucleon and that the dominant contribution comes from the one-nucleon pole diagram

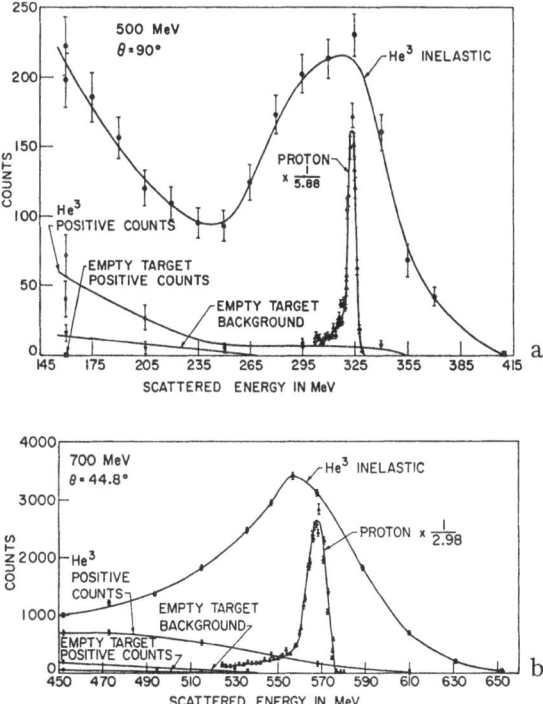

Fig. 16 a and b. Inelastic continua for He³ including comparison proton peaks

shown in Fig. 18. The further simplifying assumption is made that the two spectator nucleons recoil in a relative S-state. Thus only the contribution from the bound deuteron in the $^3S_1$ state is kept and the relative momentum of the two nucleons in the $^1S_0$ state is neglected. This treatment is similar to that given by DURAND [13] in his consideration of the electrodisintegration of the deuteron.

The cross section corresponding to the diagram shown in Fig. 18 is

$$\frac{\mathrm{d}^2\sigma}{\mathrm{d}E_f\,\mathrm{d}\Omega_f} = \frac{\sigma_{NS}}{(2\pi)^3}\left\{\frac{3}{2}\,\gamma_A^2\,g_1\,G_p(q^2) + \left[\frac{1}{2}\,\gamma_B^2\,G_p(q^2) + \gamma_C^2\,G_n(q^2)\right]g_0\right\} \quad (13)$$

where

$$\sigma_{NS} = \frac{\alpha^2\cos^2\dfrac{\theta}{2}}{4E_i^2\sin^4\dfrac{\theta}{2}}\,\frac{E_f}{E_i} \quad (14)$$

3*

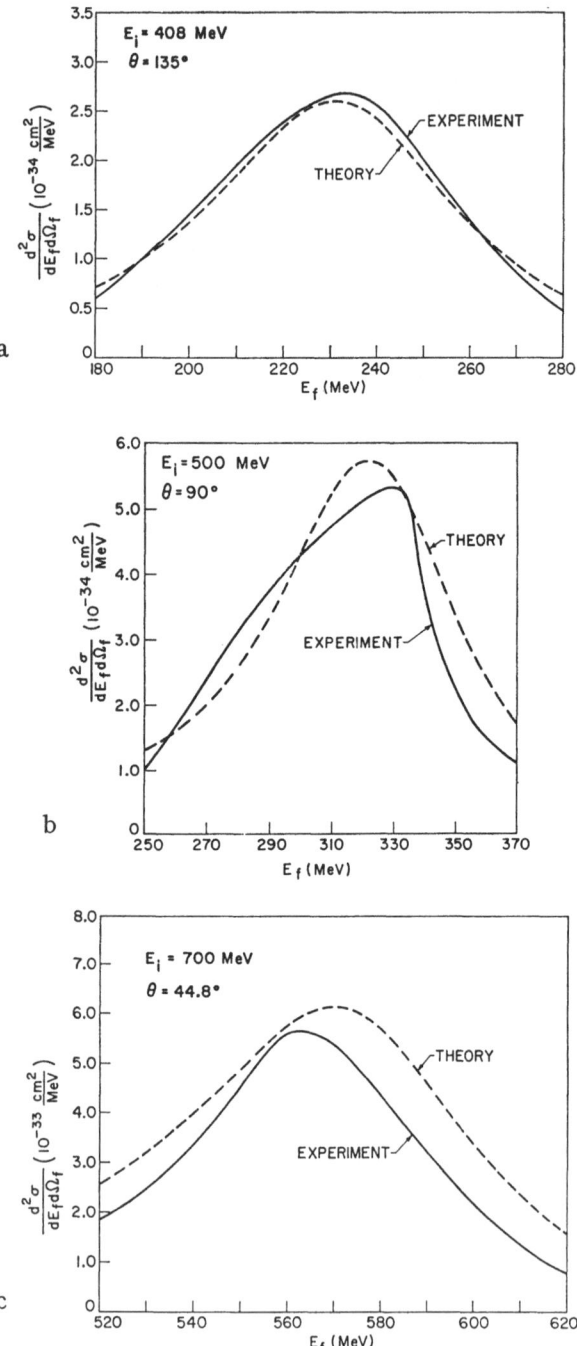

Fig. 17a—c. Experimental curves for He³ corrected as indicated in the text compared with theoretical curves of Oakes and Griffy

and

$$G_i(q^2) = \frac{G^2_{E_i}(q^2) + \frac{q^2}{4M^2} G^2_{M_i}(q^2)}{1 + \frac{q^2}{4M^2}} + \frac{q^2}{2M^2} G^2_{M_i}(q^2) \tan^2 \frac{\theta}{2} . \qquad (15)$$

The symbols have their usual meanings and M is the mass of the nucleon. E_i and M_i refer to electric and magnetic properties of the nucleon.

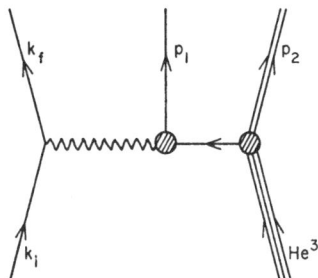

Fig. 18. A Feynman graph for the most important process calculated by OAKES and GRIFFY in the electro-disintegration of He³

The major energy dependence of the cross section is given by the functions g_0 and g_1, where the subscripts refer to the spin of the two-nucleon system. In the pole approximation these functions are of the form

$$g(E_i, E_f, \theta) = M_1 M_2 \int \frac{d^3 p_1}{E_1} \int \frac{d^3 p_2}{E_2} \times$$
$$\times \frac{\delta^{(3)}(\vec{p}_1 + \vec{p}_2 - \vec{q}) \, \delta^{(1)}(E_1 + E_2 - \omega - M_3)}{[(p_1 - q)^2 - M_1^2]^2} . \qquad (16)$$

The constants γ_A, γ_B and γ_C which occur in Eq. (13) are related to the reduced widths for the breakup of He³ into d + p, (n + p)$_{J=0}$ + p and (p + p)$_{J=0}$ + n, respectively. M_1, M_2 and M_3 refer to the masses of the similarly labeled momenta in Fig. 18. The energy transfer in the lab system is ω.

The results shown in Fig. 17 compare rather well with the experimental data, considering their present crudity and the various corrections that need to be made. Note that there are *no* free parameters in the calculations. The constants $\gamma_A = 105 \text{ MeV}^{\frac{1}{2}}$, $\gamma_B = 115 \text{ MeV}^{\frac{1}{2}}$ and $\gamma_C = 121 \text{ MeV}^{\frac{1}{2}}$ were taken from a similar analysis of the electron-proton coincidence cross sections in H³ and He³ (reference [20]) and the nucleon form factors were taken from reference [1]. It is interesting that the positions of the peaks and the half widths are nearly right in Figs. 17a and b. The discrepancies are larger in Fig. 17c but the general behavior is still correct. Both the calculations and the experiments can undoubtedly be improved but the present theoretical approach is very encouraging*.

3. Magnetic octupole moments in light nuclei. Recently an investigation of magnetic scattering has been conducted at STANFORD by studying

---

* See note added in proof on page 42.

the elastic scattering of electrons at 180° [21]. At exactly 180° the electric charge scattering vanishes almost completely and only the

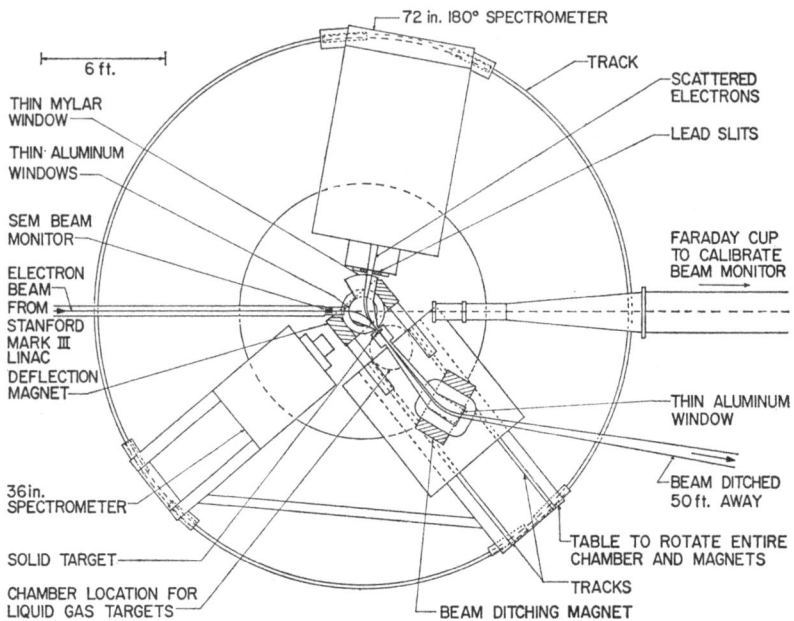

Fig. 19. A diagram of the 180° scattering apparatus

magnetic scattering appears. Because of a finite solid angle and other corrections a small residual charge scattering must be taken into account.

The equations are given below

$$\frac{d\sigma}{d\Omega} = \left(\frac{d\sigma}{d\Omega}\right)_C + \left(\frac{d\sigma}{d\Omega}\right)_M \tag{17}$$

$$\left(\frac{d\sigma}{d\Omega}\right)_C = \left(\frac{Ze^2}{2E_0}\right)^2 \frac{\cos^2\dfrac{\theta}{2} + \dfrac{1}{\gamma^2}\sin^2\dfrac{\theta}{2}}{\eta_1(\theta)\sin^4\dfrac{\theta}{2}} F_C^2(q^2) \tag{18}$$

$$\left(\frac{d\sigma}{d\Omega}\right)_M = \left(\frac{e^2}{2M_Nc^2}\right)^2 \frac{1+\sin^2\dfrac{\theta}{2}}{\eta_1^2(\theta)\sin^4\dfrac{\theta}{2}} \left(\frac{J+1}{3J}\right)\mu_0^2 F_M^2(q^2) \tag{19}$$

in which

$$F_M^2(q^2) = \left[F_{M1}^2(q^2) + \frac{2q^4}{1575}\left(\frac{2J+3}{2J-1}\right)\left(\frac{J+2}{J-1}\right)\left(\frac{\Omega_0}{\mu_0}\right)^2 F_{M3}^2(q^2) + \cdots\right] \tag{20}$$

$$\eta_1(\theta) = 1 + \frac{2E_0}{Mc^2}\sin^2\frac{\theta}{2} \tag{21}$$

and where γ is the ratio of incident energy to electron rest mass. M_N is the mass of the nucleon and μ_0 is the nuclear magnetic dipole moment in nuclear magnetons. F_{M1} is the dipole distribution form factor and F_{M3} the octupole form factor. The quantity Ω_0 is the static magnetic octupole moment of the nucleus.

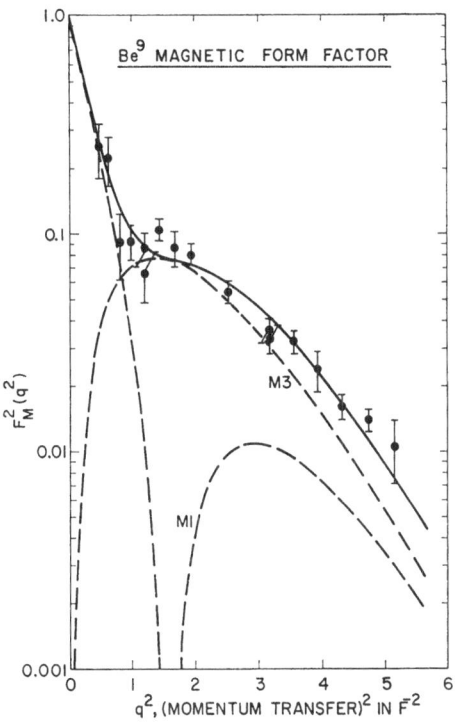

Fig. 20. This figure shows the 180° elastic scattering data in Be⁹ and the theoretical interpretation in terms of dipole and octupole contributions

The permitted multipolarity, λ, is less than or equal to $2J$, where J is the nuclear spin. The charge scattering requires even values of λ, and the magnetic scattering odd values of λ.

Near $\theta = \pi$, it may be shown that

$$\frac{d\sigma}{d\Omega} \cong \left(\frac{Ze^2}{2E_0}\right)^2 \frac{F_0^2(q_\pi^2)}{\eta_1(\pi)} \left[\frac{\xi_0^2}{4} + g(\Delta\Omega) + \frac{\langle\theta^2\rangle_{MS}}{2} + \frac{\langle\theta^2\rangle_B}{4} + \frac{1}{\gamma^2}\right] + \\ + \left(\frac{e^2}{2M_Nc^2}\right)^2 \frac{2}{\eta_1^2(\pi)} \left(\frac{J+1}{3J}\right) \mu_0^2 F_M^2(q_\pi^2) \tag{22}$$

where $(\pi - \xi_0)$ is the average scattering angle, $g(\Delta\Omega)$ is a quantity depending on the solid angle geometry, $\langle\theta^2\rangle_{MS}$ is the effective mean square multiple scattering angle in the target, and $\langle\theta^2\rangle_B$ is the mean square divergence of the incident beam.

The experiments were carried out with the geometry shown in Fig. 19 which employs an ancillary circular magnet. The 72″ spectrometer is used to receive and resolve the elastically scattered electrons from all others.

The form factors for magnetic scattering can be calculated for a harmonic well shell model for P-shell nuclei, with the following results

$$F_{M1}(q^2) = (1 - K x)\, e^{-x}\, [\text{C. M.}]\, F_N(q^2) \tag{23}$$

$$F_{M3}(q^2) = e^{-x}\, [\text{C. M.}]\, F_N(q^2) \tag{24}$$

where

$$x = \frac{q^2 a_0^2}{4} \tag{25}$$

and [C. M.] is a center of mass correction factor, and $F_N(q^2)$ is the usual nucleon form factor. The quantities K and (Ω_0/μ_0) are dependent on the type of coupling chosen in the nucleus. The quantity a_0 is a parameter related to the curvature of the harmonic well [22].

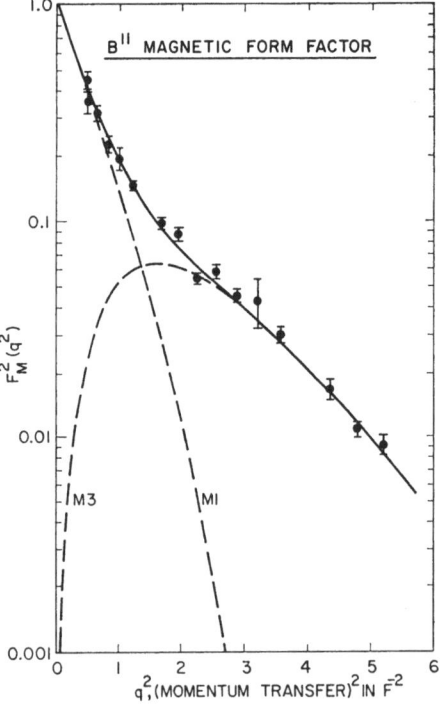

Fig. 21. This figure is similar to Fig. 20 and shows data for B^{11} along with a theoretical interpretation in the form of a solid line

Experimental data are shown in Fig. 20 for recent data taken on Be^9, a spin $J = 3/2$ nucleus. The dipole and octupole contributions are readily apparent. The harmonic well model can be fitted very well and a value for the radius of the well is obtained in agreement with previous charge scattering results. The values obtained for K and (Ω_0/μ_0) show that $L - S$ coupling is slightly preferred over other types of coupling. Finally we show in Fig. 21 results for B^{11} which is also a spin $3/2$ nucleus. The data show a preference for the $j - j$ coupling scheme.

D. Shower Studies at 900 MeV

Investigations of electron showers are being made at 900 MeV in copper and lead. These studies are carried out by Mrs. C. CRANNELL

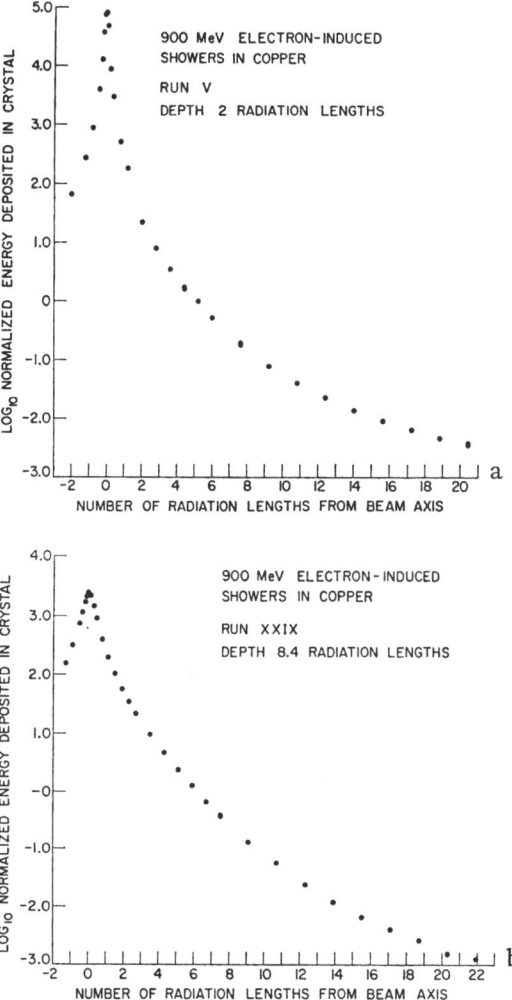

Fig. 22a and b. Shower curves in copper at 900 MeV. The data represent radial distributions at the indicated longitudinal depths in the shower

who uses a technique similar to that employed previously by KANTZ and HOFSTADTER [23]. In this technique the energy loss is measured by a DC scintillation detector mounted on the end of a long probe. The probe is placed in a "probe plate" and this plate can be moved to any longitudinal position in the shower. Radial energy loss curves are obtained directly by moving the probe in a radial direction.

Some examples of energy loss curves are shown in Fig. 22. From a set of these curves the longitudinal development of a shower can be traced. This is now being done.

Acknowledgment. The authors wish to thank the staff of the Stanford High Energy Physics Laboratory for their gracious help in operating the linear accelerator. We are also greatly indebted to Dr. George L. Rogosa of the Atomic Energy Commission for obtaining the Ca^{44} target for our use. The special skill of the Target Fabrication Group at Oak Ridge National Laboratory is especially acknowledged for preparing the Ca^{44} target in the form of a metal.

Note added in proof: The result given in Equation (13) for the pole term contributions to the cross section for the electrodisintegration of He^3 is incorrect in that σ_{NS} should be replaced by $\sigma_{\text{Mott}} = (E_i/E_f)\,\sigma_{NS}$. This change effects primarily the normalization of the theoretical curves shown in Fig. 17 and makes the theoretical results 30 to 70 per cent high if one uses the same values for γ_A, γ_B and γ_C.

References

[1] Hughes, E. B., T. A. Griffy, M. R. Yearian, and R. Hofstadter: Phys. Rev. **139**, B 458 (1965)

[2] Janssens, T., E. B. Hughes, M. R. Yearian, and R. Hofstadter: (to be published in the Physical Review)

[3] Dudelzak, B., G. Sauvage, and P. Lehmann: Nuovo cimento **28**, 18 (1963)

[4] Drickey, D. J., and L. N. Hand: Phys. Rev. Letters **9**, 521 (1962)

[5] Lehmann, P., R. Taylor, and R. Wilson: Phys. Rev. **126**, 1183 (1962)

[6] Buchanan, C. D., and M. R. Yearian: (to be published)

[7] Erickson, E.: Ph. D. Thesis, Stanford University (1965)
   Adler, R. J.: Ph. D. Thesis, Stanford University (1965)

[8] Benaksas, D., D. Drickey, and D. Frerejacque: Phys. Rev. Letters **13**, 353 (1964)

[9] Jankus, V. Z.: Phys. Rev. **102**, 1586 (1956)

[10] Adler, R. J., and S. Drell: Phys. Rev. Letters **13**, 349 (1964)

[11] Hughes, E. B., M. R. Yearian, and R. Hofstadter: (to be published)

[12] Blankenbecler, R.: Phys. Rev. **111**, 1684 (1958)

[13] Durand III, L.: Phys. Rev. **123**, 1393 (1961)

[14] Croissiaux, M., R. Hofstadter, A. E. Walker, M. R. Yearian, D. G. Ravenhall, B. C. Clark, and R. Herman: Phys. Rev. B **137**, 865 (1965)

[15] Hahn, B., D. G. Ravenhall, and R. Hofstadter: Phys. Rev. **101**, 1131 (1956)

[16] Suelzle, L. R., and M. R. Yearian: Nucleon Structure, p. 360 (edited by R. Hofstadter and L. I. Schiff. Stanford University Press 1964)

[17] Clark, B. C., R. Herman, R. Hofstadter, G. Nöldeke, K. van Oostrum, D. G. Ravenhall, and M. R. Yearian: (to be published)

[18] Collard, H.: (to be published)

[19] Oakes, R. J., and T. A. Griffy: (to be published)

[20] Griffy, T. A., and R. J. Oakes: (to be published in Rev. Mod. Phys.)
   Johansson, A.: Phys. Rev. B **136**, 1030 (1964)

[21] Rand, R. E., R. Frosch, and M. R. Yearian: Phys. Rev. Letters **14**, 234 (1965)

[22] Hofstadter, R.: Ann. Rev. Nucl. Sci. **7**, 231 (1957). See p. 272

[23] Kantz, A., and R. Hofstadter: Phys. Rev. **89**, 607 (1953)
   — — Nucleonics **12** (3) 36 (1954)

Prof. Dr. Robert Hofstadter

Department of Physics, Stanford University
Stanford/California (USA)

Review of Nucleon Form Factors

Richard Wilson

Harvard University, Cambridge, Mass.

3 years ago, before the CEA began operation, I made a survey [1] of the data then available on electron proton and electron deuteron scattering and the form factors derived therefrom. At that time 3 laboratories were in the field. Now we have 5 — Stanford, Cornell, Orsay, CEA and DESY. It is interesting that most of the data available in 1962 has been replaced in the meantime by more accurate data.

Analysis of data is now simpler. I will discuss all data in terms of the electric and magnetic form factors G_E and G_M which were introduced three years ago. We then have the famous Rosenbluth formula

$$\frac{d\sigma}{d\Omega} = \frac{(\alpha\,r_0)^2}{\left(2E\sin\dfrac{\theta}{2}\right)^2}\,\frac{E'}{E}\left\{\left(\cot^2\frac{\theta}{2}\right)\frac{G_E^2(q^2) + \tau G_M^2(q^2)}{1+\tau} + 2\tau\,G_M^2(q^2)\right\}$$

$$\tau = q^2/4M^2\,.$$

At the request of several theorists I will present data with the units of q^2 in $(\text{BeV}/c)^2$. For ease of memory $1\ (\text{BeV}/c)^2 = 25\ \text{fermi}^{-2}$.

We can now list the useful data.

From Stanford

1) 63 electron proton scattering cross sections measured by JANSSENS et al. [2] from $q^2 = 0.16$ to $1.2\ (\text{BeV}/c)^2$ to an accuracy of about 4%.

2) 39 ratios of quasi-elastic electron-neutron to electron-proton scattering cross sections measured by HUGHES et al. [3] from $q^2 = 0.29$ to $1.36\ (\text{BeV}/c)^2$ to about 6% accuracy.

From Stanford and Harvard Collaboration

3) 8 electron proton scattering cross sections from $q^2 = 0.012$ to $0.09\ (\text{BeV}/c)^2$ measured by DRICKEY and HAND [4]. These are now the

most precise e-p scattering cross sections measured with absolute accuracies of 1%. Also measured by DRICKEY and HAND are elastic electron deuteron scattering cross sections in the same range.

From Cornell

4) 21 electron-proton scattering cross sections measured by BERKELMAN et al. [5] to about 10% accuracy.

5) 10 ratios of quasi-elastic electron-neutron to electron-proton scattering measured by AKERLOF et al. [6] from $q^2 = 0.43$ to 1.36 $(BeV/c)^2$ to about 10% accuracy.

6) 4 ratios of quasi-elastic electron-neutron to electron-proton scattering cross sections measured by STEIN et al. [7] at 0.19 and 0.39 $(BeV/c)^2$ using electron-neutron and electron-proton coincidences to 10% accuracy.

From Orsay

7) 23 e-p scattering cross sections measured by LEHMANN et al. [8] from $q^2 = 0.06$ to 0.4 $(BeV/c)^2$ to about $2^1/_2\%$ absolute and 1% relative accuracies.

8) Some e-p scattering cross section measured by D. FREREJACQUE [9] from $q^2 = 0.01$ to 0.19 $(BeV/c)^2$.

9) 6 e-d elastic scattering cross sections, measured by D. BENAXAS et al. [10] from $q^2 = 0.11$ to 0.19 $(BeV/c)^2$.

From Harvard

10) 7 e-p scattering cross sections, using proton detection by DUNNING et al. [11] to about 11% accuracy.

11) 8 electron-proton scattering cross sections measured to an accuracy of 6% by DUNNING et al. Also 18 ratios of quasi elastic e-n/e-p ratios measured by electron detection and 4 e-p/e-n ratios measured by an anticoincidence method. These give a greater statistical precision.

12) One new e-p scattering cross section using the new Harvard external beam, highly tentative and for propaganda use only, at $q^2 = 0.39$ $(BeV/c)^2$ to a hoped for 2% accuracy.

13) 15 e-p scattering cross sections from DESY [13] from $q^2 = 0.39$ $(BeV/c)^2$ to 1.8 $(BeV/c)^2$ to 10% accuracy.

14) The neutron-electron scattering data from Columbia and the Argonne to 3% accuracy [14].

15) There are rumours, about which ZICHICHI will talk later in the conference, of measurements from CERN on the reaction $P + \overline{P} \rightarrow$ $\rightarrow e^+ + e^-$ and $P + \overline{P} \rightarrow \mu^+ + \mu^-$ at $q^2 = -7$ (BeV/c)^2. So far I only have upper limits available.

16) There have been measurements at Stanford [16] on the comparison of $e^+ + P$ scattering to $e^- + P$ scattering from $q^2 = 0.1$ to $q^2 = 0.73$ (BeV/c)^2. Measurements have been made up to $q^2 = 2.4$ (BeV/c)^2 at Harvard but not yet analysed. At $q^2 = 0.73$ (BeV/c)^2 and $\theta = 90°$, the cross section ratio $\sigma_{e^- P}/\sigma_{e^+ P} = 0.93 \pm 0.03$. This is barely significant and we will call it unity.

17) Measurements at ORSAY [17] give polarization of the recoil proton in e-p scattering at $q^2 = 0.6$ (BeV/c)^2 indicating also the validity of the first Born approximation.

Old disagreements between laboratories have now vanished and new smaller ones have taken their place. Unfortunately this is partly due to the effort we have all made to make the systematic errors the same in all the laboratories.

In order to compare data taken at different energies and angles at different laboratories I will compare it with certain models, assuming first Born approximation.

There are various ways of making such a comparison of models with experimental data. One is to say that Harvard data have shown there are no cores and the low momentum transfer data tells us the r. m. s. radii. Then we try a fit and let the rest of the data fall where they will.

Another way is to take all 200 pieces of data and to adjust parameters on a computer for the best fit. The two methods of approach give almost the same results, but since the second method is more intimidating, and shows that we have worked hard, I will present the results from this method. Dr. JOHN DUNNING has done most of this work.

LEVINGER and PEIERLS extrapolated electron-proton scattering cross sections to the time-like region. They found that there must be a resonance around 700 MeV, but they found that much more detailed statements were not possible. I will here reverse the procedure and assume resonances and see how they fit.

In order to understand the neutron form factors I must extract them from the data on the deuteron. I here choose the inelastic electron deuteron scattering data. I will assume that the impulse approximation is valid for $q^2 > 0.3$ (BeV/c)^2 and ignore data below. The formula is given by DURAND and is

$$\frac{d^2\sigma}{d\Omega\,dE_3} = \left[\left(\frac{d\sigma}{d\Omega}\right)_P + \left(\frac{d\sigma}{d\Omega}\right)_N\right] \frac{E_1}{E_3} \frac{P_{cm}}{E_{cm}} M(p, q) .$$

Integrating over the quasi-elastic peak one finds approximately

$$(d\sigma/d\Omega)_{e\text{-}d} = (d\sigma/d\Omega)_P + (d\sigma/d\Omega)_N .$$

The kinematic function $M(p, q)$ is known from the knowledge of the deuteron ground state wave function. The final state interaction correction is small and has been discussed by NUTTALL and WHIPPMANN.

GOURDIN has a paper at this meeting where he discusses these relativistically. He believes that data below $q^2 < 0.3$ $(BeV/c)^2$ can now be correctly understood. However, he limits the applicability of his arguments to $q^2 < 1$ $(BeV/c)^2$ because of effects of the first resonance. It is hard to see how this could affect the conclusion in the integrated e-d cross section.

It was at the Aix-en-Provence conference in 1691 that R. R. WILSON suggested use of G_E and G_M following a suggestion of L. HAND. P. LEHMANN and I realized that the first precise e-p points from Orsay fitted the simple relation $G_{MP} = G_{EP}$. This relation was, of course, scoffed at by theorists except KEN BARNES who thought of it independently. It is interesting that it is a good fit to all the data to write

$$G_{EP} = \frac{G_{MP}}{\mu_P} = \frac{G_{MN}}{\mu_N} = \frac{4M}{q^2} \frac{G_{EN}}{\mu_N}$$

$$= \frac{1}{(1 + q^2/0.71)^2} \quad \text{when } q^2 \text{ is measured in } (BeV/c)^2$$

$$= \frac{1}{(1 + q^2/18.1)^2} \quad \text{when } q^2 \text{ is measured in fermi}^{-2}.$$

In 1957 at a conference in Stanford, HOFSTADTER suggested this last line as a fit to the data — it gives a Fourier transform $e^{-\lambda r}$. You will remember that he then measured only backward angle points and hence G_{MP}.

Theorists now find the relations

$$G_{EP} = \frac{G_{MP}}{\mu_P} = \frac{G_{MN}}{\mu_N}$$

$$G_{EN} = 0$$

are a consequence of $SU(6)$. Now at the threshold for the proton anti-proton annihilation $q^2 = -4M^2$ and $G_E = G_M$ unless F_1 and F_2 are infinite. Either these relations are approximate, or else $G_E = G_M = 0$.

This fit I call "fit 1" for future reference and give it a name, the dipole fit. The mass of the dipole is at a value $\sqrt{0.71} = 0.85$ BeV, for instead of a simple pole at the mass 0.85 BeV it corresponds to two poles close together and with opposite signs of coupling.

It has been conventional to try to fit the data by the meson resonances of spin 1 and negative parity. The $\rho(I = 1)$ ω and $\Phi(I = 0)$ are now known. Unfortunately when the ω and ρ masses were first known in 1961, it was already clear (at the Aix-en-Provence conference where FUBINI tried it) that this simple picture would not work.

We have tried to fit the data using ω and Φ mesons at the correct mass and the ρ meson at a mass shifted to fit the data. If we insist $G_E = G_M$ at $q^2 = -4M^2$ we must include cores also. We find a bad fit to the Harvard data. In particular the e-p cross section is higher than the fit by a factor of 3 from $q^2 = 3$ to $q^2 = 7$ $(BeV/c)^2$. Fig. 1 shows the χ^2 as a function of the variable ρ mass.

Our best, but still lousy, fit with 3 poles and cores is

$$G_{EV} = -0.025 + \frac{0.525}{1 + q^2/(0.5)^2}$$

$$G_{ES} = -0.035 + \frac{1.43}{1 + q^2/(0.78)^2} - \frac{0.93}{1 + q^2/(1.02)^2}$$

$$G_{MV} = -0.118 + \frac{2.44}{1 + q^2/(0.5)^2}$$

$$G_{MS} = -0.040 + \frac{1.06}{1 + q^2/(0.78)^2} - \frac{0.59}{1 + q^2/(1.02)^2}$$

which differs little from a fit by HUGHES [3] to less data.

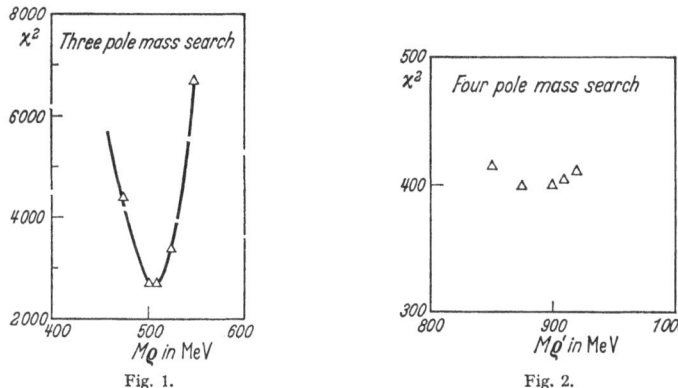

Fig. 1.   Fig. 2.

Fig. 1. The goodness of fit parameter χ^2 for a 3 pole mass search. The expected χ^2 is 200

Fig. 2. The goodness of fit parameter χ^2 for a 4 pole mass search. The expected χ^2 is 200

The trouble with the 3 pole fit is that for the isovector form factors we have failed to approximate the dipole. For the isoscalar form factors, the ω at lower mass, and the Φ at higher mass, approximate it well. So we can now try to use a ρ' meson at a variable mass and put the ρ meson at its correct mass.

This fit I will call "fit 4". Fig. 2 shows the effect of varying the ρ' mass. We keep the ω and Φ mesons at their correct masses. This fit has parameters

$$G_{EV} = \frac{2.27}{1 + q^2/(0.76)^2} - \frac{1.77}{1 + q^2/(0.9)^2}$$

$$G_{ES} = \frac{1.18}{1 + q^2/(0.78)^2} - \frac{0.68}{1 + q^2/(1.02)^2}$$

$$G_{MV} = \frac{7.37}{1 + q^2/(0.76)^2} - \frac{5.01}{1 + q^2/(0.9)^2}$$

$$G_{MS} = \frac{1.05}{1 + q^2/(0.78)^2} - \frac{0.61}{1 + q^2/(1.02)^2} \; .$$

Figs. 3, 4, 5, and 6 show the fits to form factors. We should note that the computer tried to fit to cross sections, so the fits to form factors look artificially good. Figs. 7 and 8 show the fits to the e-p and e-n cross sections at high momentum transfers and show how badly fit 3 fits the Harvard data.

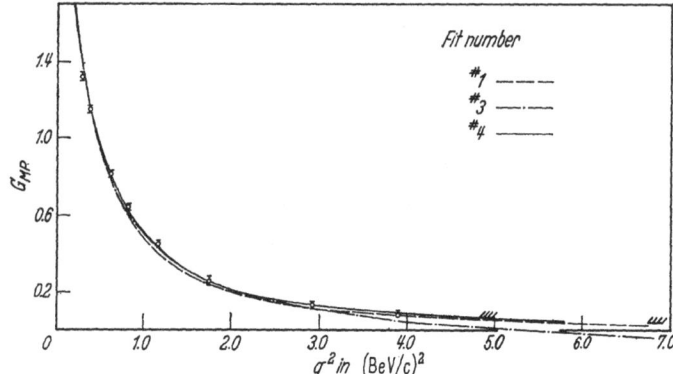

Fig. 3. The form factor G_{Mp} compared with 3 of the models of the text

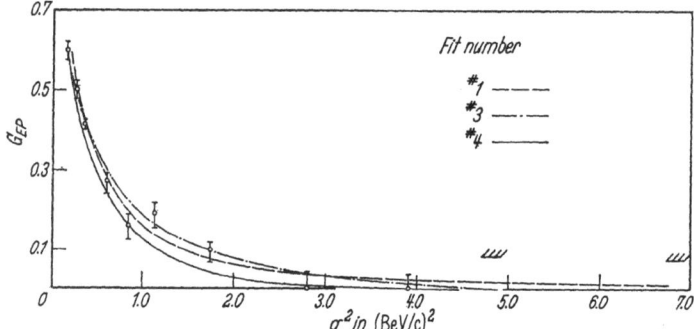

Fig. 4. The form factor G_{Ep} compared with 3 of the models of the text

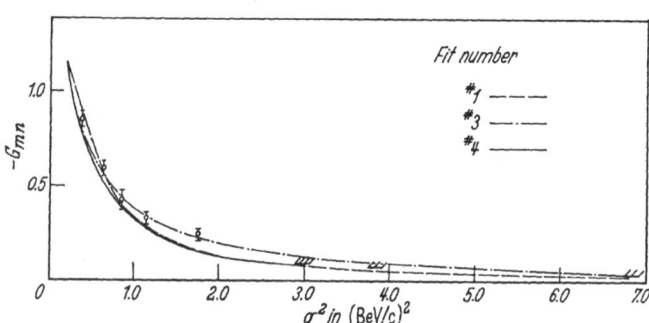

Fig. 5. The form factor G_{Mn} compared with 3 of the models of the text

With these fits as interpolations between data points of different laboratories I can now point out discrepancies between different laboratories. I note at the start that if a fit were made with the data of one laboratory only, we could get $\chi^2 = 1.1$ per point; yet adding laboratories, all laboratories contribute their proportionate share to χ^2 and no laboratory is perfect.

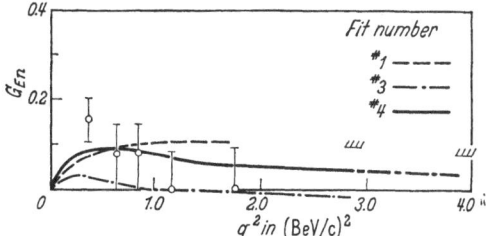

Fig. 6. The form factor G_{En} compared with 3 of the models of the text

1) Janssens' backward angle e-p points are systematically lower than those of LEHMANN which are already lower than those of HAND.

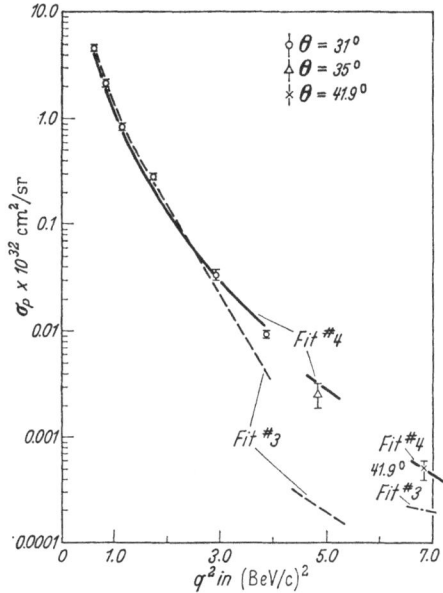

Fig. 7. The electron proton differential cross section compared with 2 of the models of the text

2) Fig. 9 shows another problem. The data of $q^2 = 0.39$ $(BeV/c)^2$ is now the most extensive test of the Rosenbluth formula. Taken by itself it suggests a value of G_E which is greater than G_M/μ_P and much greater than any of the fits 3 and 4. The points that suggest this are the small angle Harvard points measured by proton detection 2 years ago [11] and our new point. Maybe this shows greater complexity to come. It might be an exchange of more than one photon with a combined angular momentum greater than 1. This can give a term in $\cot^4\theta/2$ which will appear as $\cot^2\theta/2$ becomes greater.

3) The neutron electric form factor is in a mess. Part of this can be solved by the new calculations of GOURDIN. But the Harvard data at

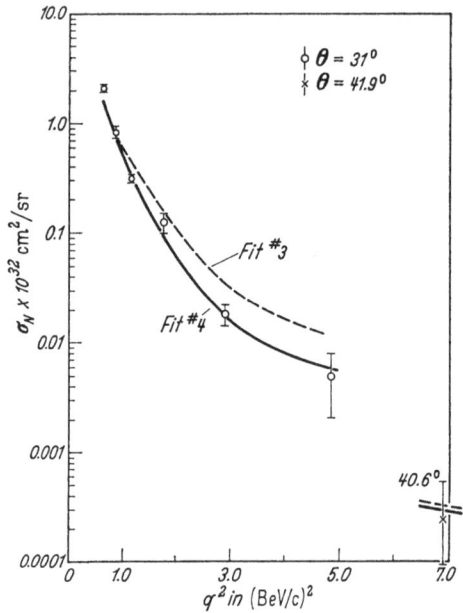

Fig. 8. The electron neutron cross section compared with 2 of the models of the text

$q^2 = 0.39$ (BeV/c)^2 and 0.59 (BeV/c)^2 is particularly disquieting because of its accuracy.

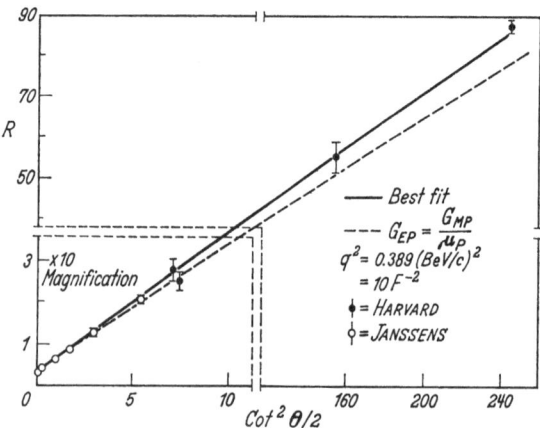

Fig. 9. The curly bracket of the first equation plotted against $\cot^2 \theta/2$ to show the G_E, G_M separation

4) The Cornell e-p data are lower than JANSSENS at 135° and higher than Harvard at 90°. I believe they are wrong.

5) The recent DESY data seems to give smaller cross sections at 50° angles than the Harvard data around $q^2 = 1$ (BeV/c)². Their data is better fitted by our theoretical lines than Harvard's.

6) The elastic electron deuteron cross sections [4, 10] when combined with the nonrelativistic deuteron form factor give $G_{EN} = 0 \pm 0.02$ up to $q^2 = 0.19$ (BeV/c)². Yet our fit gives $G_{EN} = 0.05$. The deuteron form factor is well determined from the bending energy, the triplet effective range, and the assumption of one-pion exchange for nuclear forces. Various theoretical guesses on the correct deuteron relativistic wave function give corrections that make G_{EN} negative accentuating the discrepancy.

7) The fits assume one-photon exchange. The Stanford value for $\sigma_{e^-p}/\sigma_{e^+p}$ of 0.93 ± 0.03 is disquieting. If one-photon exchange is not correct the fits may not be meaningful.

Now we have to ask, do these parameters seem reasonable? Firstly no ρ' meson has been found. The symmetry theorists do not want one. Worse, still, experiments which should have found it if it exists have not. In particular there is an experiment to be reported to this conference, in which mesons are photoproduced by a diffraction process. If ρ' mesons couple to γ-rays they should also be photoproduced in the same way, but they have not been found.

This is not definitive and we shall have to wait for data on the leptonic decays of vector mesons or experiments with colliding electron and positron beams.

Since what we really want is a dipole for the ρ meson, we can reduce the mass of the ρ meson and increase that of the ρ' without altering the fit.

The coupling constants for the ω and Φ are maybe easier to understand. Low would like the ω-coupling to be reduced by the A quantum number. The Φ couples only weakly to nucleons. So the products $f(\omega\,\gamma)\,f(\omega\,N\,N)$ may be about equal to $f(\Phi\,\gamma)\,f(\Phi\,N\,N)$.

Next we can compare nucleon form factors to two predictions from strong interactions.

Firstly we take the idea of WU and YANG [19], who noted that p-p scattering varies as $\exp(-p_\perp/0.15)$. Since for e-p scattering there is one interacting cloud and for p-p scattering there are 2 interacting clouds e-p scattering varies less fast than this. They suggest $G(q^2) = \exp(-q^2/0.6)$. Fig. 10 shows the fit. At low momentum transfers it is bad — and indeed has the wrong analytic behaviour. We can do better by putting $G(q^2) = \dfrac{1}{1.5}\exp\left(-\sqrt{q^2 + 4m_\pi^2/0.6}\right)$ which has the right analytic properties and about the right radius 1.2 fermis whereas the correct one is 0.8 fermis.

This fit corresponds to *many* resonances.

In the time-like region $G(q^2) = \exp(i\sqrt{-q^2}/0.6)$ which gives an alternating value, corresponding to resonances every 1.8 BeV with alternating phase. This is then the opposite extreme from the fits with limited number of poles we described earlier.

4*

For ZICHICHI's work on $P + \overline{P} \to e^{+} + e^{-}$ these two fits give $G^2 = 1/100$ and 1 respectively. If one remembers that one cannot really distinguish in the space-like region $\exp(i\sqrt{-q^2}/0.6)$ from a set of sharp resonances, the possibilities are greater. But averaged over a range of antiproton energies Wu's fit suggests $G^2 = 1$.

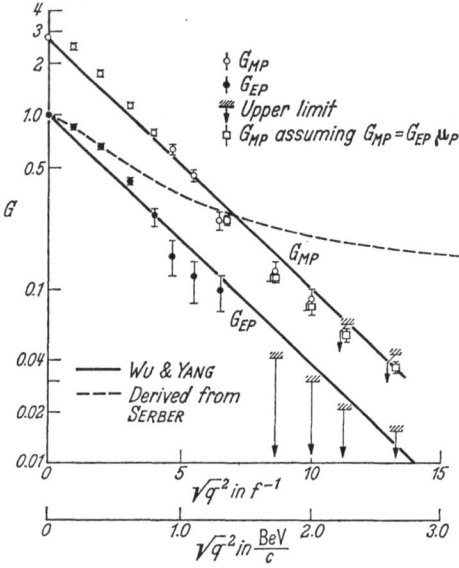

Fig. 10. The nucleon form factors compared with the equation of Wu and Yang

Secondly, a naive idea of mine suggested by an optical analysis of P-P data by SERBER. He finds that some of the main features can be explained by means of an imaginary potential

$$V(r) = \frac{i\hbar c}{r} \exp(-\lambda r)$$
$$\lambda = 2.12 \, f^{-1}.$$

If we attribute a charge distribution to this potential $\varrho(r) \propto V(r)$.

$G(q^2) = \lambda^2/(q^2 + \lambda^2)$ and if we allow for the fact that in P-P scattering we have two interacting clouds we get $G = [\lambda^2/(q^2 + \lambda^2)]^{1/2}$ as shown in the dotted line of the figure. We get about the *right* r. m. s. radius (0.65 fermi) but the wrong shape.

Elastic scattering merely measures the transition between two nucleon states:

$$\langle N \, |j_\mu| \, N \rangle$$

Clearly the transitions to excited nucleon states N are equally interesting:

$$\langle N \, |j_\mu| \, N^* \rangle.$$

PANOFSKY and ALLTON began such a study in 1957. We get a cross section for inelastic electron scattering which can be separated into longitudinal and transverse parts corresponding to the G_E and G_M separation.

The transverse cross section $\sigma_\perp$ should have a simple behaviour depending on the multipolarity of the transition.

$$\sigma_\perp \sim [G_{MV}(q^2)]^2 |\vec{q}^*|^{2l} \qquad \text{magnetic multipoles}$$

$$\sim [G_{MV}(q^2)]^2 |\vec{q}^*|^{2l-2} \quad \text{electric multipoles}$$

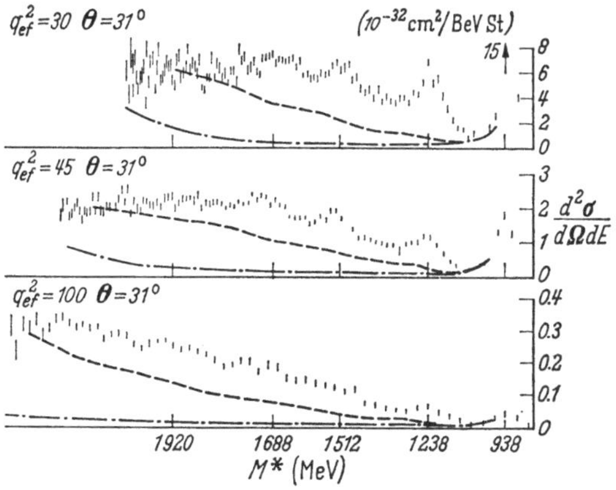

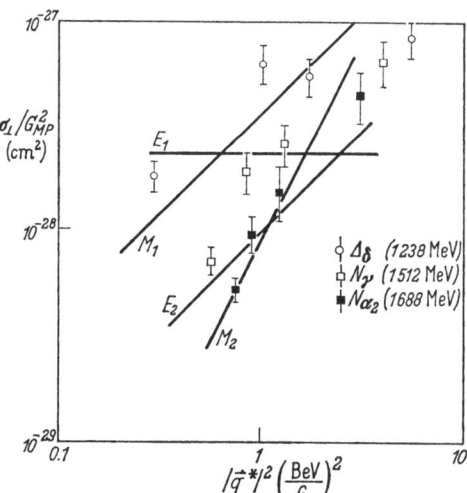

Fig. 11. The spectrum of electrons scattered inelastically from protons

and equals σ_γ as $q^2 \to 0$. $\vec{q}^*$ is the 3 dimensional momentum transfer in the excited nucleon rest system. We note by comparison that the

Fig. 12. The transverse cross section plotted against $(\vec{q}^*)^2$ to show the effect of multipolarity

equivalent cross section for elastic scattering varies as $q^2 G_M^2(q^2)$. Fig. 11 shows inelastic electron scattering data showing excitation of the first 3 nucleon resonances [20]. The dotted lines are calculations of the radiative rail. However, I must confess that these calculations are incorrect. We have found a mistake. The corrections are less than shown. Fig. 12 shows $\sigma_\perp$ derived from the data of Fig. 11, and photoproduction data, plotted against $|\vec{q^*}|^2$. The solid lines are the expected threshold dependencies. There is agreement for the 1238 MeV resonance (magnetic dipole) and the 1688 MeV resonance (magnetic quadrupole) but the 1512 MeV resonance gives the wrong behaviour. It looks like magnetic quadrupole whereas we except electric dipole. I think this is a fruitful field which is just beginning.

References

[1] Hand, L. N., D. G. Miller, and Richard Wilson: Rev. Mod. Phys. 35, 335 (1963)
[2] Janssens, T.: Thesis. Stanford 1964
[3] Hughes, E. B., T. A. Griffy, M. R. Yearian, and R. Hofstadter: Phys. Rev. 139, B 458 (1965)
[4] Drickey, D., and L. N. Hand: Phys. Rev. Letters 9, 521 (1962)
[5] Berkelman, R., M. Feldman, R. M. Littauer, G. Rouse, and R. R. Wilson: Phys. Rev. 130, 2061 (1963)
[6] Akerlof, C. W., K. Berkelman, G. Rouse, and M. Tigner: Phys. Rev. 135, 810 (1964)
[7] Stein, P., R. W. McAllister, B. D. McDaniel, and W. M. Woodward: Phys. Rev. Letters 9, 403 (1962) and also private communication
[8] Lehmann, P., R. Taylor, and Richard Wilson: Phys. Rev. 126, 1183 (1962)
[9] Frerejacque, D.: Thesis, Paris (1965)
[10] Benaxas, D., D. Drickey, and D. Frerejacque: Phys. Rev. Letters 13, 353 (1964)
[11] Dunning, J. R., K. W. Chen, N. F. Ramsey, J. K. Rees, W. Shlaer, J. K. Walker, and Richard Wilson: Phys. Rev. Letters 10, 500 (1963)
[12] Dunning jr., J. R., K. W. Chen, A. A. Cone, G. Hartwig, N. F. Ramsey, J. K. Walker, and Richard Wilson: Phys. Rev. Letters 13, 631 (1964)
[13] Hartwig, G., F. Brasse, H. Schopper: (DESY)
[14] Melkonian, E., B. M. Rustad, and W. W. Havens: Phys. Rev. 114, 1571 (1959)
[15] Zichichi, A.: Private communication
[16] Browman, A., F. Liu, and C. Schaerf: Phys. Rev. Letters 12, 183 (1964)
[17] Bizot, J. C., J. M. Buon, J. Lefrancois, J. Perez, y Jorba, and Ph. Roy: Phys. Rev. Letters 11, 480 (1963)
[18] Hofstadter, R., F. Bumiller, and M. R. Yearian: Rev. Mod. Phys. 30, 482 (1958)
[19] Wu, T. T., and C. N. Yang: Phys. Rev. 137, B 708 (1965)
[20] Cone, A. A., K. W. Chen, J. R. Dunning, G. Hartwig, N. F. Ramsey, and Richard Wilson: Phys. Rev. Letters 14, 326 (1965)

Prof. Dr. R. Wilson

Physics Department
Harvard University of Cambridge
Cambridge 38, Mass.

Special Models and Predictions for Pion Photoproduction (Low Energies)

G. Höhler

Institut für Theoretische Kernphysik
der Technischen Hochschule Karlsruhe

1. Introduction

The theoretical investigations on pion photoproduction can be classified into two groups; the phenomenological analysis and the attempts to treat the dynamics of the $\pi N \gamma$ system.

In the phenomenological analysis [1] only the general theoretical principles are used, namely

a) Lorentz and gauge invariance (including space and time reflection).

b) The principle of minimal electromagnetic interaction, which states that the electromagnetic interaction has to be introduced by $p_\mu \rightarrow p_\mu - eA_\mu$. It leads to a relation between the photoproduction processes in different charge states.

c) The unitarity condition. Together with the time reversal invariance it allows one to deduce the "final state theorem", according to which the phase of a multipole amplitude is equal to the scattering phase shift in the final state, if the energy is below the inelastic threshold

$$M_{lJ} = \pm |M_{lJ}| \, e^{i \delta_{lJ}} . \tag{1}$$

In b) and c) the electromagnetic field is treated only to lowest order.

The aim of the phenomenological analysis (or multipole analysis) is analogous to that of the phase-shift analysis in πN scattering. One tries to determine the multipole amplitudes from the experimental data, since these amplitudes are much better suited for a comparison with the predictions of a dynamical theory than the cross sections.

A dynamical theory which allows one to calculate the photoproduction amplitudes from first principles does not yet exist. In recent years all attempts to predict the amplitude were based on the dispersion relation approach [2—4], which has a more modest aim. One tries to calculate the photoproduction amplitudes from some consequences of the axioms of field theory [5] together with the experimental information from other reactions, as for instance π N and ep scattering.

At present the dispersion relation approach is not a systematic theory. In order to obtain a prediction one has to make drastic approximations, which were found using the static model of CHEW and LOW as a guide [2]. There is no reliable way to estimate the errors.

Omitting all indices the dispersion relation at fixed t for the production amplitude reads

$$\operatorname{Re} A\,(s,\,t) = A_{\text{pole}}\,(s,\,t) + \frac{1}{\pi}\,P \int\limits_{(M+1)^2}^{\infty} \mathrm{d}s'\,\operatorname{Im} A\,(s',\,t)\left\{\frac{1}{s'-s} \pm \frac{1}{s'-\bar{s}}\right\} \quad (2)$$

where $s = W^2$, $W =$ total energy in the c. m. system, $t =$ invariant momentum transfer squared, $\bar{s} = -s - t + 2M^2 + 1$, $m_\pi = 1$, $M =$ mass of the nucleon.

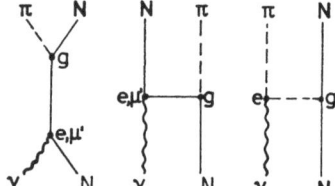

Fig. 1. Feynman graphs of the Born terms

The pole term follows from the one-nucleon intermediate states. It is also called the "Born term", since it happens to agree with the amplitude calculated from the Feynman graphs of Fig. 1, if the pseudoscalar π N coupling is used ($f^2 = g^2/4M^2 = 0.081$) and the electromagnetic coupling of the anomalous moment μ' of the nucleon is taken into account explicitly.

The most important contribution to $\operatorname{Im} A\,(s',\,t)$ in the dispersion integral is expected to belong to the isobar intermediate state

$$\pi + \mathrm{N} \to \Delta \to \pi + \mathrm{N}$$

where Δ denotes the 33-resonance at the total c. m. energy $W = 1236$ MeV. This transition can proceed via the magnetic dipole amplitude M_{33} and the electric quadrupole amplitude E_{33} only. If all the other contributions to $\operatorname{Im} A$ are neglected and $\operatorname{Re} A$ is calculated from Eq. (2), the result will be called in the following the "isobar approximation".

2. The Isobar Approximation

2.1. The resonant multipoles. The first problem in all investigations using the isobar approximation is to find an expression for the energy dependence of the resonant multipoles M_{33} and E_{33}.

CGLN [2] derived approximate "dispersion relations" for the resonant partial waves by a projection of the fixed-t dispersion relations, neglecting the non-33 contributions to the dispersion integral and considering the static limit and some recoil corrections. They found that the "dispersion relations" for the resonant π N partial wave f_{33} and the resonant multipole M_{33} agree, if the formula

$$M_{33} = \frac{\mu_v}{f} \frac{k}{q} f_{33}; \quad f_{33} = \frac{e^{i\alpha_{33}} \sin \alpha_{33}}{q} \tag{3}$$

is assumed to hold and only a certain part of the Born term of M_{33} is taken into account, namely the static limit of the anomalous magnetic part + the recoil correction of order $1/M$ to the electric part. In (3) $\mu_v = (\mu_p - \mu_n)/2$ denotes the isovector part of the total magnetic moment of the nucleon, k the photon momentum and q the pion momentum in the c. m. system*. CGLN also gave an estimation of the remaining part of M_{33} and of E_{33}, assuming that these multipole amplitudes have a similar behaviour as in the static model. The contribution of these terms is small in comparison with (3).

Several authors have tried to improve the result of CGLN. Recently FINKLER [7] used the Omnès method, assuming that f_{33}, M_{33} and E_{33} have the same real phase at all energies. His further assumption that all non-33 contributions to the dispersion integrals can be neglected has to be discussed critically, since DONNACHIE and HAMILTON [8] had found an appreciable $T = 0$ π π-contribution in their investigation of the resonant f_{33} amplitude. FINKLER corrected Eq. (3) by a factor on the right hand side, which is about 0.9 at the resonance and decreases at higher energies, similar to McKinley's result [9]. Furthermore he obtained a large correction to the CGLN estimate of the E_{33}/M_{33} ratio, which is of special importance for the π^0 production with polarized γ-rays [10]. A detailed discussion has not yet been made, but it seems that the discrepancy found in earlier calculations will be reduced considerably by Finkler's treatment**.

2.2. The work of Ball and Schmidt. BALL [11] approximated Im A by the contribution of Im M_{33} alone, taking M_{33} from Eq. (3). Then he evaluated the dispersion integrals and calculated the cross sections without further approximations. Unfortunately his numerical results are not reliable, because there is an error in his Eq. (8.29) and also in the D-coefficient for π^0-production.

Ball's work was continued by W. SCHMIDT [12–14], who calculated predictions for all measured quantities up to 500 MeV, except for the polarization of the recoil proton in π^0 production, which was treated by MÜLLENSIEFEN [15].

Since f^2 and α_{33} were taken from the scattering data, the prediction is an absolute one, it contains no adjustable parameters.

---

* Other derivations of Eq. (3) are discussed in § 9.3 of ref. [1] and § 2.2.3 of ref. [6].

** Note added in proof. W. SCHMIDT and H. WUNDER have shown that the prediction agrees well with the experimental α/C-data below 350 MeV.

A comparison with the experimental data shows that the Ball-Schmidt calculation leads to a reasonable zero order approximation for the photoproduction cross sections, including the measurements with polarized γ-rays and the π^-/π^+ ratios. However one should notice that the agreement is mainly due to the fact that the cross sections are dominated by the pole terms and the resonance effects. From the experience with the isobar approximation in πN scattering* one would expect that the Ball-Schmidt result could be quite wrong for some of the small multipoles, especially for those leading to $T = 1/2$ final states.

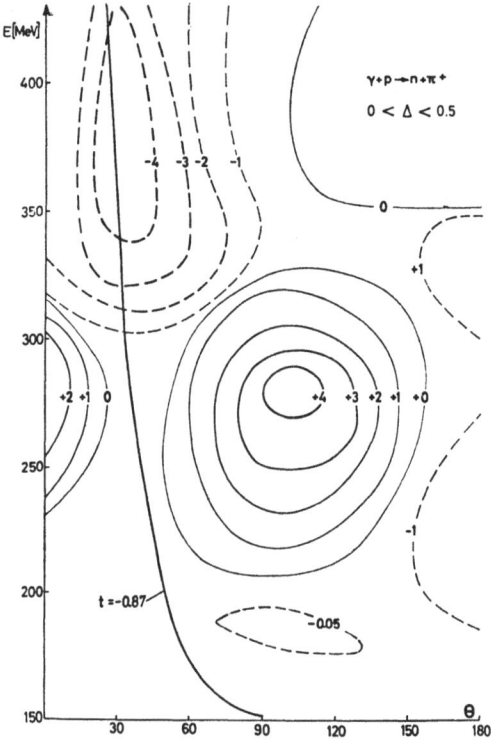

Fig. 2. Contour diagram of the difference between the experimental differential cross sections and Schmidt's prediction in μb/st. The curve denoted by $t = -0.87$ gives the relation between energy and angle, if the squared 4-momentum transfer t is equal to its value at threshold

Fig. 2 shows a contour diagram of the difference between Schmidt's prediction for π^+ production and the experimental data, interpolated by a Moravcsik fit. The deviation amounts to 15% at its maximum near 280 MeV. In the region of small angles the extrapolation of the experimental data is only a crude estimate and the magnitudes of the deviations will be much better known after the completion of the new measurements at Orsay and Bonn.

---

* See Fig. 6 of ref. [16].

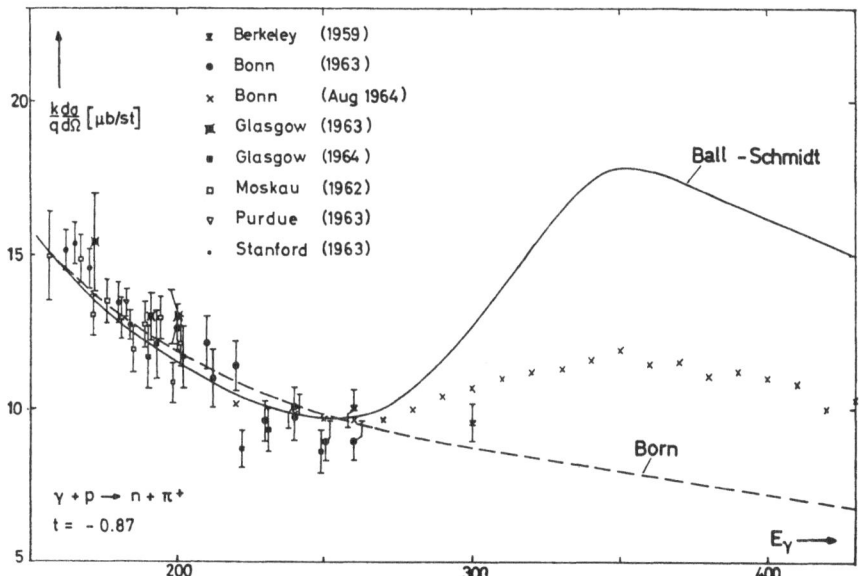

Fig. 3. π^+-cross section for a momentum transfer equal to its threshold value (cf. Fig. 2)

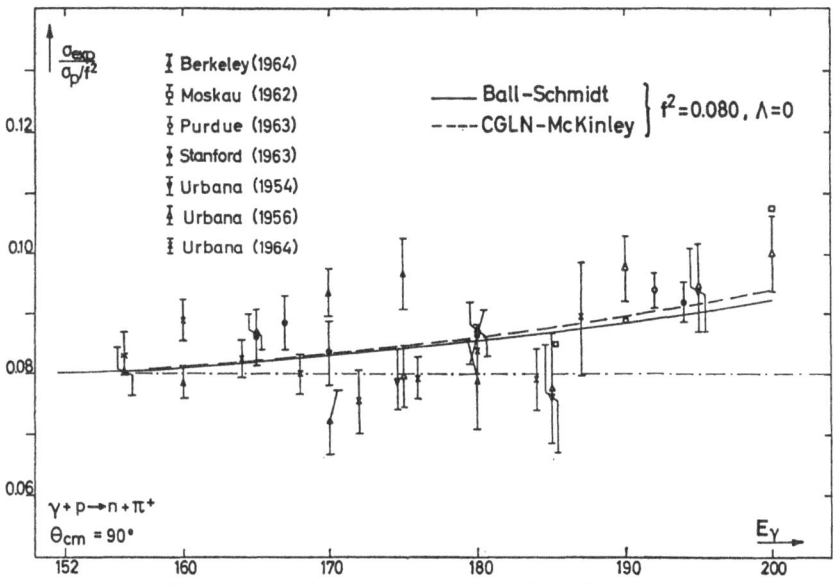

Fig. 4. π^+-cross section at 90° c.m. near threshold σ_p = Born cross section

BALDIN [17] has pointed out that it is interesting to consider the cross section at a fixed momentum transfer t which is equal to its value at threshold, since in this case the integrand of the dispersion integral (2) is used in the physical region of (s', t) only. In all other cases there is an additional uncertainty following from the extrapolation of $\mathrm{Im}\, A\,(s',t)$ into the unphysical region. According to Fig. 3 the agreement between the Ball-Schmidt prediction and the experimental data is good up to 260 MeV. It is interesting to notice that in this region the data can be as well described by the cross section calculated from the pole term alone

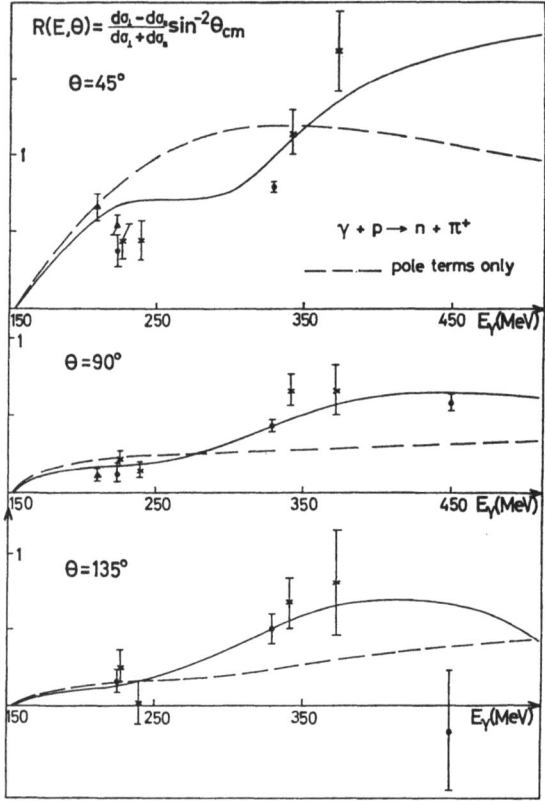

Fig. 5. π^+ excitation curves for plane-polarized γ-rays

($=$ second order Born approximation). At higher energies there is an increasing and rather large deviation (cf. the curve $t = -0.87$ in Fig. 2) which must be due to an error of $\mathrm{Im}\, A$ in the physical region.

The error caused by the extrapolation into the unphysical region should be especially large for the excitation curve at 180°, however Fig. 2 shows that in this case the Ball-Schmidt prediction agrees very well with the experiments.

Although the recent π^+ production data at 90° near threshold are very accurate, they do not allow one to test the dispersion integral contribution, since the Born term is so much larger (Fig. 4). Unfortunately one cannot use these data for an accurate determination of the coupling constant f^2, since one expects several other slowly energy dependent contributions from the dispersion integral, which cannot be estimated in a reliable way. The same difficulty is present, if one wants to test the well-known relation between the Panofsky ratio, the difference of the πN s-wave scattering lengths and the photoproduction data [18].

The experimental data on π^+ production with polarized γ-rays are compared with the prediction in Fig. 5.

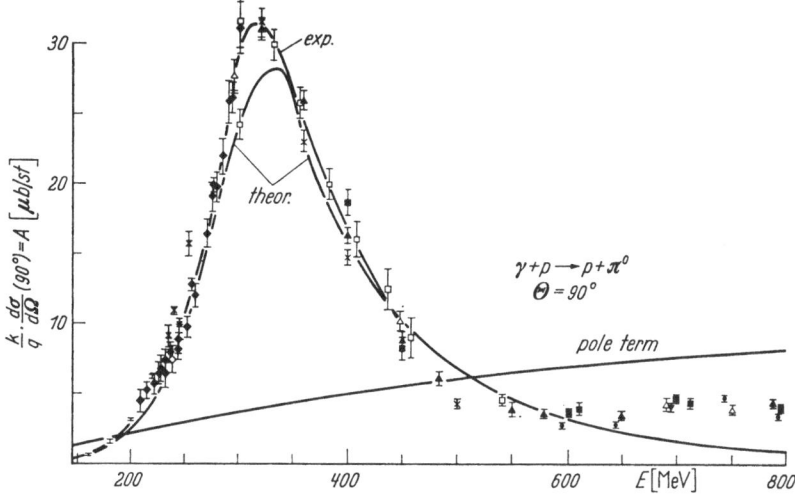

Fig. 6. Excitation curve for π^0-production at 90° c.m.s.

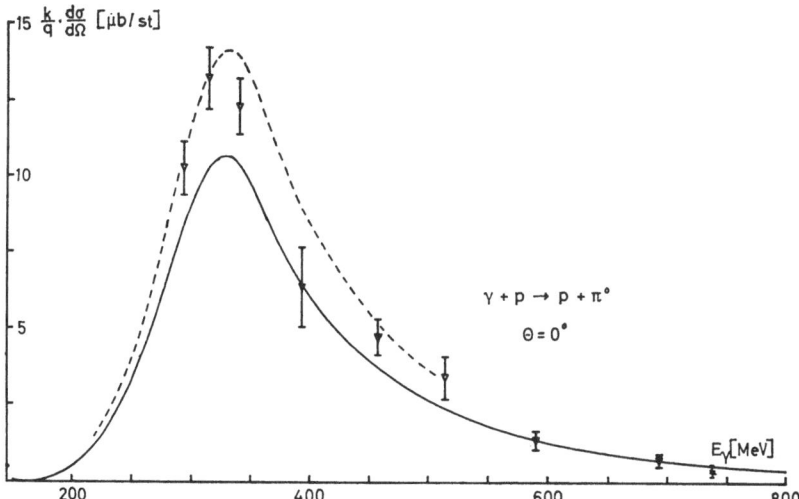

Fig. 7. Excitation curve for π^0-production at 0°. Dashed line: estimation of the correction from $\mathrm{Im}\,M_{33} \times \mathrm{Im}\,E_{0+}$

The experimental information on π^0-production is not as good as for π^+-production. Figs. 6, 7, 8 show that the Ball-Schmidt prediction again is a zero order approximation. However it is easier to find large deviations, since the Born term is smaller than for π^+ production and partly compensated by an indirect effect of the resonance. Therefore the cross sections are more sensitive to the "small" multipoles, for which the isobar approximation is not reliable. For instance there is a large discrepancy [13] in the energy dependence of B and C ($\sigma = A + B\cos\theta + C\cos^2\theta$) below 300 MeV and in the ratio α/C, which follows from the experiments with polarized γ-rays [10].

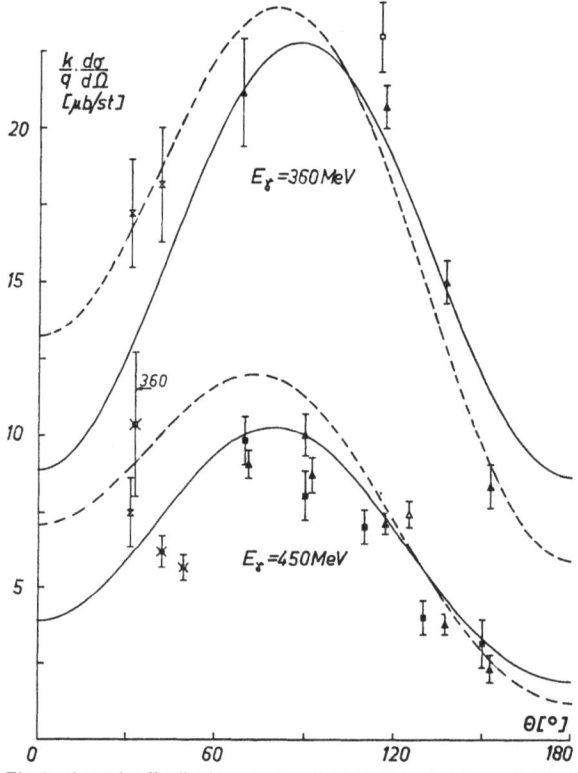

Fig. 8. π^0-angular distributions at 360 and 450 MeV. Dashed line as in Fig. 7

It is astonishing that the simple Ball-Schmidt calculation describes so many features of pion photoproduction, although it *does not contain adjustable parameters*. If one discusses the comparison with the experimental data, one should keep in mind that appreciable corrections are expected from several other contributions to the amplitude, but at present they cannot be calculated in a reliable way. The most interesting experiments are those which show deviations from the prediction far outside the errors, since they help to identify those parts of the theory, which are in need of improvement.

The work of Schmidt is only the simplest version of the isobar approximation. It should be improved by taking into account the unitarity condition for the 33-multipoles in a better way. First steps in this direction have already been made by CGLN [2], BALL [11] and by McKINLEY [9]. It will be interesting to see, to what extent FINKLER's careful treatment [7] of the unitarity condition for the resonant multipoles diminishes the discrepancies found by SCHMIDT*.

2.3. Feynman Graphs. AMATI and FUBINI [19] have pointed out that in the limit of a narrow resonance the isobar approximation of the dispersion integral leads to the same result as the evaluation of the Feynman graphs of Fig. 9. The first term in the integrand of Eq. (2) corresponds to the graph I and the second term to II.

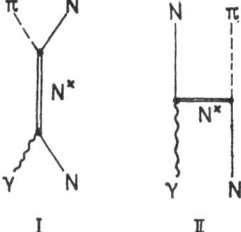

Fig. 9. Feynman graphs for isobar intermediate states

It is interesting to notice that the $1/(s' - s)$ term in the dispersion integral gives a large contribution to $\mathrm{Re}\,E_{0+}^{(+)}$ and thereby to a $J = 1/2$ final state [18]. For the corresponding graph I in Fig. 9 this is somewhat unexpected, since the intermediate isobar state has the spin 3/2.

If one starts from the usual interaction Lagrangian one finds an additional term in the interaction Hamiltonian which leads to the $J = 1/2$ final state**.

GOURDIN and SALIN [21] and RASHID and MORAVCSIK [22] treated the isobar intermediate states in such a way that the graph I does not contribute to $J = 1/2$ final states. It is not clear to me how these calculations can be justified from the general theoretical principles.

The comparison between Schmidt's calculation and the Feynman graph formulas gives the values of the $\gamma\,NN^*$ coupling constant, the $\pi\,NN^*$-coupling constant following from a similar treatment of $\pi\,N$ scattering [18]. Furthermore it shows that the description of the finite width by a constant imaginary part of the isobar mass $(M^* + i\,\Gamma)$ is not sufficient for quantitative purposes. If one wants to use a Breit-Wigner type formula one has to assume a strongly energy-dependent width***.

3. Corrections to the Isobar Approximation

3.1. Final-state corrections to the non-33 multipoles. In the isobar approximation the unitarity condition is not fulfilled for the non-33 multipoles. An estimation of the final-state corrections was given by CGLN [2]. It was improved by taking into account relativistic kinematics in the paper of McKINLEY [9], but the theoretical derivation is still more or less doubtful. Also the addition of these corrections to the

---

* See the footnote on p. 57.

** I am much indebted to Prof. H. UMEZAWA, Prof. A. VISCONTI and Dr. G. VON GEHLEN for discussions on this question. Also I would like to thank Dr. HADJIOANNOU for sending me a preprint which treats a closely related aspect of this problem [20].

*** Cf. the interesting discussion of the isobaric model in the review article of GELL-MANN and WATSON [23].

isobar approximation amplitude does not lead to a better agreement with the experiments*.

In my opinion one does not gain much information, if the experiments are compared with a prediction which contains many uncertainties. It is better to determine the multipoles by a phenomenological analysis (§ 4) and to compare each multipole with the theoretical expression, as given for instance in the paper of McKinley.

3.2. The ρ-exchange contribution. In the isobar approximation the dispersion integral does not contribute to the isoscalar part A_i^0 of the

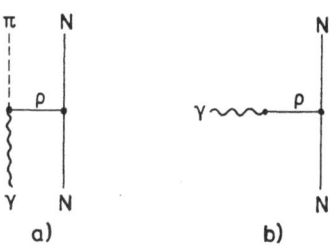

a)   b)

Fig. 10. Graphs for ρ-exchange processes

amplitude. But if one considers the fixed-s dispersion relation [11] or the Cini-Fubini approximation to the Mandelstam representation [24], one is led to expect a contribution from the ρ-meson exchange in the t-channel (Fig. 10a). Ball [11] succeeded in expressing the result by the isovector part of the electromagnetic nucleon form factor which is assumed to be dominated by the ρ-exchange effect (Fig. 10b).

If an empirical fit to the experimental form factors and the new data for the ρ-resonance are inserted into Ball's result, one finds for the ρ-exchange contribution to the amplitude [18] (we give A_1^0 only)

$$A^0_{1\rho}(s, t) = 4.1 \frac{\Lambda}{e} \mu'_v \left[0.50 + \frac{0.60\,t}{18 - t} \right]. \tag{4}$$

For energies in the region of the first resonance or below, t is so small that the first term in the bracket is dominating and a multipole decomposition shows that the main contributions belong to E_{0+} and M_{1-}. It will be very difficult to distinguish the ρ-exchange parts from other corrections to these multipoles (for instance from final-state interactions or high energy contributions to the dispersion integrals) as long as one considers π^+- or π^0-production directly. The situation is more favorable, if the data are combined in such a way that the isoscalar part is isolated or enhanced (π^-/π^+ ratio). At present there is no convincing evidence for the ρ-exchange effect [18]. Ball's coupling constant Λ is smaller than $0.5\,e$, unless the ρ-exchange is masked by another correction to the isobar approximation.

Gourdin et al. [24] have also noticed the relation to the nucleon form factor, but instead of evaluating the first term in the bracket of Eq. (4) they suggested treating it as an adjustable parameter. The same suggestion was made by De Tollis et al. [25]. Since McKinley [9] neglected this term which is the dominating one in Ball's result, without giving a reason, his treatment of the ρ-exchange effect is questionable.

The values of Λ given in the experimental papers should not be compared with each other without a critical examination of the underlying

---

* Note added in proof. An interesting new investigation of this problem has been made by A. Donnachie et al. (private communication).

theoretical analysis. In several cases the results for Λ differ not only because of the experimental data but also because of different assumptions and definitions.

4. Phenomenological Analysis

4.1. Summary of the results of the phase shift analysis.
Because of the close connection between pion photoproduction and πN scattering it is useful to summarize first our knowledge of the πN system as obtained from the scattering data, which have considerably improved during the last year[*]. The quantum numbers of the 2nd, 3rd and 4th resonances are now well established and it is clear that other strong effects occur mainly in the states P_{11}, S_{11}, S_{31} (indices: $2T, 2J$)[**]. The following table gives some of the properties which are relevant for photoproduction. All energies are γ-laboratory energies, $E_\gamma = T_\pi + 150$ MeV.

δ is the real part of the phase shift, η the absorption parameter. The last column gives the energy at which the real part of the resonant scattering amplitude has its maximum. It is an estimate of the position of a peak or a dip, caused by the interference between the real part of the resonant multipole and a slowly varying real background amplitude.

Table 1.

	T	J^p	multipoles	$\delta = 90°$ E_γ (MeV)	η_{min}	$|Re f| = $ max at E_γ (MeV)
Λ (1236) 1st res.	$\dfrac{3}{2}$	$\dfrac{3}{2}^+$ P_{33}	M_{1+}, E_{1+}	345	1.0	290, 480
N (1525) 2nd res.	$\dfrac{1}{2}$	$\dfrac{3}{2}^-$ D_{13}	M_{2-}, E_{2-}	770	0.25	680
N (1680) 3rd res.	$\dfrac{1}{2}$	$\dfrac{5}{2}^+$ F_{15}	M_{3-}, E_{3-}	1040	0.6	960, 1120
Λ (1920) 4th res.	$\dfrac{3}{2}$	$\dfrac{7}{2}^+$ F_{37}	M_{3+}, E_{3+}	1495	0.2	1290, 1680
N (1400) ?	$\dfrac{1}{2}$	$\dfrac{1}{2}^-$ P_{11}	M_{1-}	≈ 750 $(\delta \approx 80°)$	0.2	490

In the discussion of the question whether the P_{11} phenomenon is a "resonance" one should keep in mind that the notion of a resonance is not sharply defined. There is a continuous transition to several other phenomena and therefore to a certain extent it is a matter of convention and of convenience how to define a resonance. For instance the

---

[*] Cf. the papers presented at the Royal Society Meeting in London (11 February 1965), to be published in the Proceedings of the Royal Society.

[**] Note added in proof. Several groups have found evidence for a D_{15}-resonance in πN scattering ($E_\gamma \approx 1050$ MeV, M_{2+}, E_{2+}) (Oxford Conference, Sept. 1965).

properties of a resonance are considerably changed if a threshold is nearby or if there is a large background in the same partial wave. Also in many models a resonance occurs together with strong variations in other partial waves and it might be unsuitable to consider it separately.

It will be very interesting to see if the P_{11} phenomenon occurs in photoproduction as inconspicuously as in π N scattering or if it is enhanced for some reason as in the final state of pp scattering at small momentum transfer and high energies [26]. If the "resonance energy" is defined by $d\delta/dE = \max$ it would be at $E_\gamma \approx 570$ MeV.

Presumably the "shoulder" in the total π^+ p cross sections near $T_\pi = 750$ MeV is not caused by a resonance, but it seems [27] that an important contribution comes from a strong variation in S_{31}.

4.2. The work of Gourdin and Salin. In their well-known work on the "isobaric model" GOURDIN and SALIN [21] have described the π^+ and π^0-production data up to $E_\gamma = 800$ MeV by an ansatz which uses Breit-Wigner type formulas for the resonances, treating the coupling constants, the widths, and several background terms as adjustable parameters. This investigation was performed three years ago. In the meantime our knowledge of the π N system has considerably improved and the present status leaves little hope that the simple ansatz of GOUR-DIN and SALIN or RASHID and MORAVCSIK [22] is adequate for a *quantitative* description of photoproduction.

Of course the ansatz could be extended by admitting additional parameters for the background multipoles and introducing the important energy dependence [23] of the resonance widths Γ (Γ_{33} for π N scattering changes by a factor of two between $\alpha_{33} = 45°$ and $135°$). But there remains the question of uniqueness, which has led to so many difficulties in the simpler case of π N scattering, and has not yet been discussed in photoproduction.

4.3. The work of Schmidt, Schwidersky and Wunder. As mentioned in 2.2, the general features of pion photoproduction below 500 MeV are well described by the results of Schmidt's evaluation of the isobar approximation. Therefore it seems reasonable to consider the possibility that the exact multipole amplitudes differ only by small corrections from the multipoles of the Ball-Schmidt approximation. In order to find these corrections, SCHMIDT [28] has calculated the variation of his prediction for the cross section σ_{BS}, if the real and imaginary part of one of the s, p, or d-wave multipoles is changed by a small amount. The result is plotted at fixed energies as a function of angle. It allows easy discussion of the different possibilities for corrections of σ_{BS} which lead to a better agreement with the experimental data.

This method corresponds to a multipole analysis which is limited to sets of multipoles in the neighborhood of the Ball-Schmidt amplitudes.

The π^0- and π^+-production data near the 2nd resonance were analyzed by SCHMIDT, SCHWIDERSKY and WUNDER [29] assuming that the cross sections can be described by the Ball-Schmidt amplitude and additions (which are not necessarily small) to E_{2-}, M_{2-}, and a few other multipoles. The isobar approximation to the dispersion relation approach cannot be applied in this region, because the polynomial expansion in $\cos\theta$ of $\operatorname{Im} A (s', t)$ does not converge any more. However this is not an objection

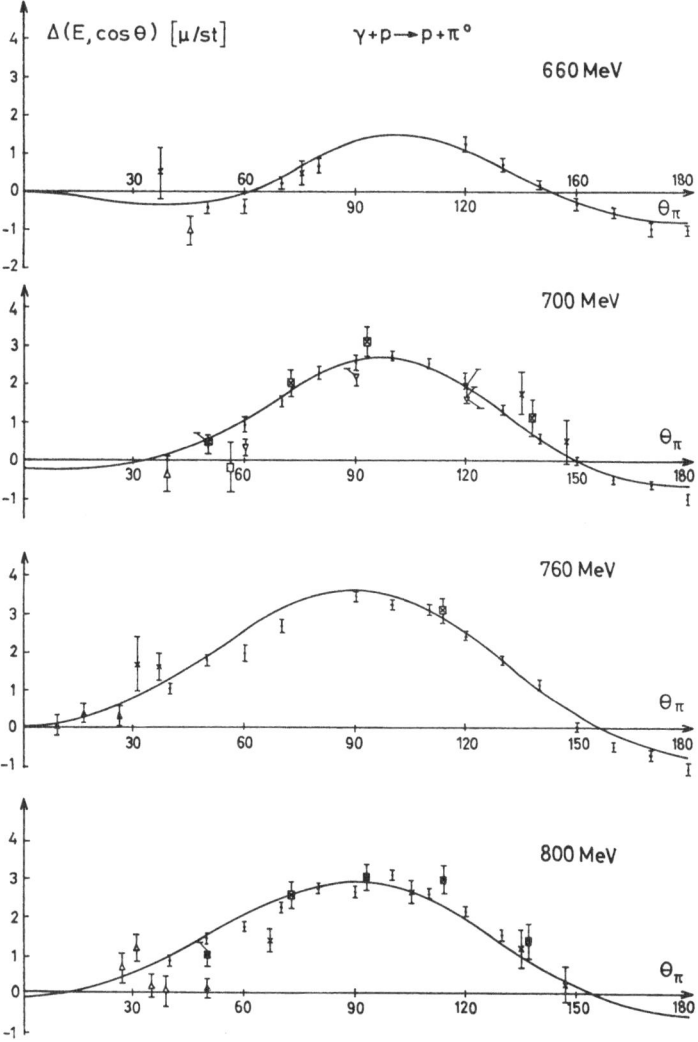

Fig. 11. π^0-production in the region of the 2nd resonance. Experimental points: Δ gives the difference between the data and Schmidt's calculation which is assumed to describe the main background effects. Solid line: best fit obtained by additional contributions in E_{2-}, M_{2-}, E_{0+}

against using the empirical fact that the extension of Schmidt's calculations to these energies reproduces the general features of the non-resonant background.

It turns out that a good fit can be obtained (Fig. 11) for a resonance-like behaviour of E_{2-} and M_{2-}. However the present data admit several solutions and there could be others which were not found because of the restricted assumptions.

The discussion of the polarization of the recoil proton in π^0-production [15, 29] has shown that a broad peak above 500 MeV is expected from the background effects alone. It would be very interesting to look for a superimposed structure at the energy of the 2nd resonance, taking into account that according to the recent results of the phase shift analysis its width is much narrower than formerly supposed [27].

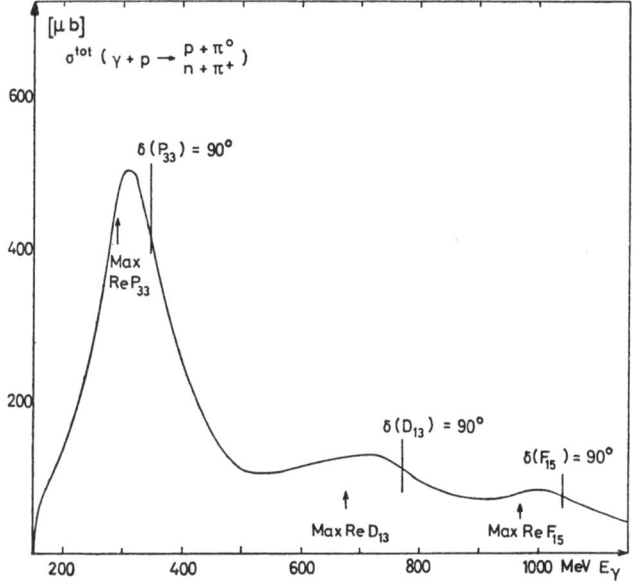

Fig. 12. Sum of total π^0- and π^+-cross sections. δ = real part of the resonant phase shift

Finally we compare in Fig. 12 the position of the resonances defined by $\delta = 90°$ with the sum of the total π^0- and π^+-cross sections. This quantity was chosen in order to eliminate all interference terms which could cause a shift of the peaks. It is seen that in all three cases there is a shift to the low energy side which presumably has to be explained for the higher resonances in the same way as for the well-known first resonance.

5. Concluding Remarks

In earlier summaries on the status of photoproduction several simple models played an important role which were not treated above, for instance the model of PEIERLS [30] which has led to the correct predictions for the quantum numbers of the 2nd and 3rd resonances.

Unfortunately there are many indications that photoproduction is more complicated than assumed in these models and cannot be described quantitatively by a small number of simple terms.

We have seen that in the low energy region there is a remarkable success of the isobar approximation to the dispersion relation approach. Although it does not contain adjustable parameters, the predictions for the absolute values of the cross sections are in most cases in reasonable agreement with the experimental data. So it seems that this approximation is a good starting point for further improvements.

At present there is no convincing evidence for the contributions of ρ- and ω-exchange processes to single-pion photoproduction. This question deserves further study since it would be very interesting to determine the coupling constants. Possibly one will encounter similar difficulties as those mentioned by Prof. VAN HOVE [31] for πN charge exchange scattering. In this case the data seem to be compatible with the assumption of the exchange of a ρ-Regge pole [32].

Presumably further progress in the theory of pion photoproduction will be made in a similar way as in pion-nucleon scattering during the last years [27]. The phenomenological analysis of the forthcoming data will give information on the energy dependence of the multipole amplitudes which can be compared with the results of a more detailed study of their dispersion relations. This does not necessarily mean that the theory will become more and more complicated, since it might be that someone will find a simple physical picture for the dominating parts of the production amplitude.

References

[1] GOLDBERGER, M. L., and K. M. WATSON: Collision Theory, § 9.2. New York: John Wiley Inc. 1964
[2] CHEW, G. F., M. L. GOLDBERGER, F. E. LOW, and Y. NAMBU: Phys. Rev. 106, 1345 (1957)
[3] LOGUNOV, A. A., A. N. TAVKHELIDZE, and L. D. SOLOVYOV: Nucl. Phys. 4, 427 (1957)
[4] SOLOVYOV, L. D.: Nucl. Phys. 5, 256 (1958)
[5] OEHME, R., and J. G. TAYLOR: Phys. Rev. 113, 371 (1959)
[6] STICHEL, P.: Fortschritte der Physik 13, 73 (1965)
[7] FINKLER, P.: Preprint, Purdue University, Lafayette, Indiana (March 1964)
[8] DONNACHIE, A., and J. HAMILTON: Phys. Rev. 133, B 1053 (1964)
[9] MCKINLEY, J. M.: Technical Report Nr. 38, Physics Department, University of Illinois, Urbana, Ill.
[10] DRICKEY, D. J., and R. F. MOZLEY: Phys, Rev. 136, B 543 (1964)
    BARBIELLINI, G., et al. (Frascati, Preprint Nov. 1964)
[11] BALL, J. S.: Phys. Rev. 124, 2014 (1961)
[12] SCHMIDT, W., and G. HÖHLER: Proc. of the Sienna International Conference 1963, p. 403
[13] — Thesis, Technische Hochschule Karlsruhe, 1964
[14] — Z. Physik 182, 76 1964)
[15] MÜLLENSIEFEN, A.: Z. Physik 188, 199, 238 (1965)
[16] HÖHLER, G., and G. EBEL: Nuclear Physics 48, 470 (1963)
[17] BALDIN, A. M.: Proc. 1960 Ann. Intern. Conf. High-Energy Physics, Rochester, p. 325, University of Rochester 1960

[18] Höhler, G., and W. Schmidt: Ann. Phys. (N. Y.) **28**, 34 (1964)
[19] Amati, D., and S. Fubini: Ann. Rev. Nucl. Sci. **12**, 359 (1962)
[20] Hadjioannou, F. T.: On the Spin 3/2 Propagator, preprint, Physikalisches Institut der Universität Bonn (June 1965)
[21] Gourdin, M., and Ph. Salin: Nuovo cimento **27**, 193 (1963)
     Salin, Ph.: Nuovo cimento **28**, 1294 (1963)
[22] Harun-ar-Rashid, A. M., and M. J. Moravcsik: preprint, Lawrence Radiation Lab., Berkeley, Cal. (UCRL-12110-T, 1965)
[23] Gell-Mann, M., and K. M. Watson: Ann. Rev. Nucl. Sci. **4**, 219 (1954)
[24] Gourdin, M., D. Lurié, and A. Martin: Nuovo cimento **18**, 933 (1960)
[25] Tollis, B. de, E. Ferrari, and H. Munczek: Nuovo cimento **18**, 198 (1960)
[26] Belletini, G., G. Cocconi, A. N. Diddens, E. Lillethun, J. P. Scalon, and A. M. Wetherell: Contribution to this conference
[27] Lovelace, C.: Invited paper at the Royal Society Meeting on Pion-Nucleon Scattering and Excited Nucleon States, to be published in the Proceedings of the Royal Society
[28] Schmidt, W.: Contribution to this conference
[29] — G. Schwidersky, and H. Wunder: Contribution to this conference
[30] Peierls, R. F.: Phys. Rev. **118**, 325 (1960)
[31] van Hove, L.: Invited paper at this conference p. 1
[32] Höhler, G., J. Baacke, J. Giesecke, and N. Zovko: Invited paper, presented at the Royal Society Meeting on Pion-Nucleon Scattering and Excited Nucleon States, to appear in the Proceedings of the Royal Society and preprint (T. H. Karlsruhe, Oct. 1965)

Prof. Dr. G. Höhler
Institut für Theoretische Kernphysik
der Technischen Hochschule
75 Karlsruhe, Hauptstraße 54

Special Models and Predictions for Photoproduction above 1 GeV*

S. D. Drell

Stanford Linear Accelerator Center
Stanford University, Stanford, California

In considering models of photoproduction processes at high energies $\gtrsim 1$ GeV, we begin with a discussion of the production of secondary beams of strongly interacting particles. These applications make the least stringent demands upon the theory since we are primarily concerned only with approximate predictions of fluxes of high energy π mesons, K mesons, or even of anti-nucleons which can then serve as projectiles in subsequent experiments. More refined quantitative results are of secondary importance for the beam production analyses and correspondingly we learn less from them about the detailed theoretical nature of the interactions involved.

In the second part of our discussion we turn to more detailed analyses of the low momentum transfer, or peripheral, processes initiated by photons. Finally in part three we consider the central, or high momentum transfer collisions.

I. Beams

The contributions from five different types of amplitudes must be included in beam production studies. These are illustrated in Fig. 1 and are expected to play the dominant roles in high energy and low momentum transfer processes, when not forbidden by selection rules.

Before discussing a new beam production process for K° mesons arising from diagram (c), I will review briefly the present situation with regard to charged pion photoproduction via one pion exchange [1] as in diagram (a) and as illustrated in more detail in Fig. 2.

* Work supported by the U.S. Atomic Energy Commission

The assumption of the peripheral calculation is that the amplitude corresponding to this graph should be the predominant one at high photon and pion energies, $k \sim \omega_q \gg \mu$ and at low momentum transfers,

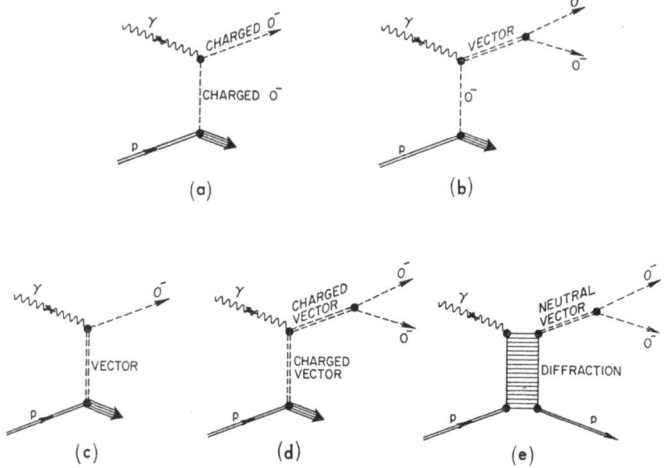

Fig. 1a—e. Amplitudes for photoproduction of meson beams

or small production angles, $\theta_q \sim \mu/k$. Under these kinematical conditions the impact parameter in the collision can be large, $\sim 1/(2\mu) = 1$ fermi, since an "almost real" pion is being exchanged between the vertices (a)

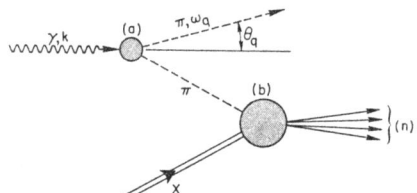

Fig. 2. One pion exchange contribution to photoproduction of a charged pion beam

and (b) in Fig. 2. The corresponding contribution to the cross section is large because the numerator factors evaluated at the pion exchange pole are large themselves, being the product of the pion current at (a) with the total pion absorption cross section at (b).

Experimental support for the approximate validity of this peripheral calculation has been established [2] for an incident photon beam with maximum energies ranging from 1.2 to 5.8 GeV as shown in the following two graphs in Fig. 3. The factor-of-two agreement near the peak of the angular distribution at $\theta = \mu/\omega_q$ supports the optimism of the peripheral model in its prediction that intense charged pion beams will be produced at electron accelerators at high energies [3]. In fact the agreement between the simple model and observation is close enough to motivate further theoretical studies and refinements aimed at accounting for their difference [4]. (See note added in proof p. 89)

Various off-mass-shell and non-pole-term corrections have been studied and diminish the difference between the model and the data.

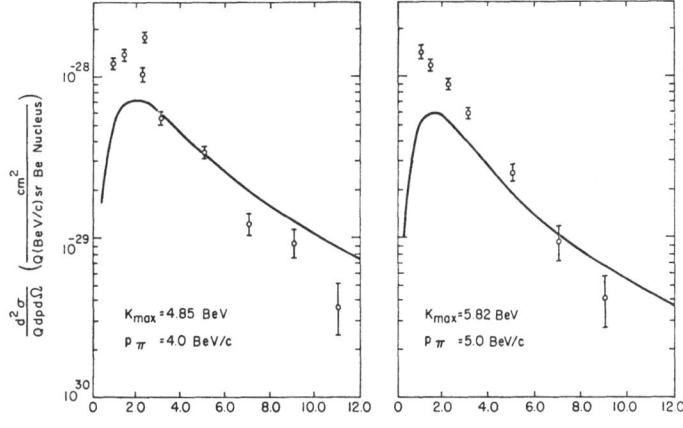

Fig. 3. Comparison of experimental results for π^- photoproduction from beryllium with predictions of the one pion exchange calculation (taken from R. BLUMENTHAL et al. [2])

In particular ITABASHI and HADJIOANNOU found that the static Chew-Low theory applied to 1.2 GeV photoproduction from hydrogen led to a

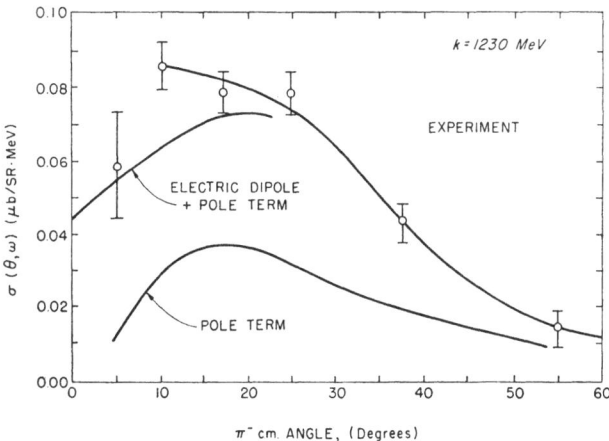

π^- cm. ANGLE, (Degrees)

Fig. 4. Comparison of calculations on $\gamma + p \rightarrow p + \pi^+ + \pi^-$ with experiment at 1.23 GeV

substantial improvement, as illustrated in Fig. 4, and recent work by STICHEL and SCHOLZ [5]*, who have included N* exchange as one

* Other models may be designed to lead to no increase. It is also very dangerous to have simple Feynman propagators for particles of spin greater than zero appearing in amplitudes for high-energy processes as they may lead to incorrect asymptotic behaviors.

possible relativistic extension of this model designed to preserve gauge invariance, leads to a similar increase. The situation is still not clear as we move in from the peak in the distribution at $\theta = \mu/\omega_q$ to the forward

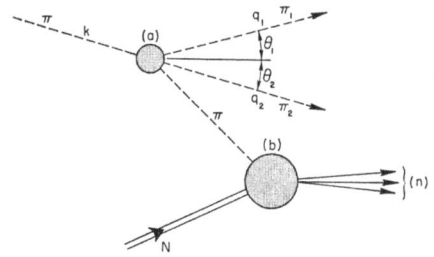

Fig. 5. One pion exchange contribution to pion production of a ρ meson

angle $\theta = 0°$ because the incident photon must transfer its angular momentum to the nucleon line. We shall return to this point later.

Before doing this, however, one can ask whether the extrapolation from 5 GeV to higher energies, say 20 GeV, can be made with any confidence on the basis of these measurements as well as of the extensive and detailed analyses of experiments with peripheral interactions of strongly interacting particles. With regard to one pion exchange processes at higher energies there is some positive evidence in the recently reported analysis [6] of peripheral ρ^0 meson production by incident pions of 12 and 18 GeV/c on carbon in the reaction $\pi^- + C \rightarrow \rho^0 + $ (anything). The relevant graph is shown in Fig. 5, and the theoretical inputs are the ρ resonance parameters (energy and decay width) at (a) and the total pion-nucleus cross sections at (b). The correlation of theory with experiment is illustrated in Fig. 6 and again illustrates a factor of two agreement. No major corrections damping this amplitude are observed or anticipated at forward directions due either to reggeization of the pion trajectory or to initial and final state absorption corrections*.

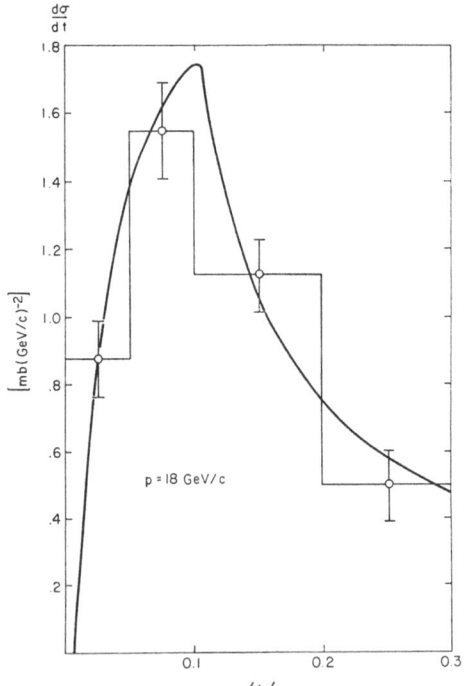

Fig. 6. Comparison of experimental results for
$\pi^- + C \rightarrow \rho^0 + $ (anything)
with predictions of the one pion exchange calculations (taken from L. Jones [6])

---

* See discussion following Eq. (1) and in footnote on page 77.

The other candidates for charged pion beam production in Fig. 1 have also been calculated [7] and only the vector current contribution in graph (d) is found to be a possible major contributer at high energies. However this contribution is significantly smaller than that of the pion current when the ρ exchange amplitude is reduced as found experimentally in the π-nucleon charge exchange analysis[8]. (See note added in proof p. 89)

How is the situation with K meson beams? Here the pole for K meson exchange is approximately $m_K \approx 500\,\mathrm{MeV}$ distant from the physical region and data [9] have not yet been obtained at sufficiently high energies so that relatively this is a small energy interval. Therefore a real quantitative test is still wanting. However the indications at present for photoproduction of K^- at 2.6 and 3.1 GeV from hydrogen are similar to the results found for the pion beam — an approximately correct magnitude ($\approx 4\,\mu$b/ster $\cdot$ BeV/c) but an angular distribution that continues to grow instead of decreasing to zero as $\theta \to 0°$.

For neutral K beams, and in particular for high energy K_2 beams, we turn to the process [10] in graph (c) of Fig. 1. There is no charge current for this process as in graph (a) and production of a charged vector K* followed by the decay

$$K^{*\pm} \to K^0 + \pi^{\pm}$$

as in (b) and (d) is calculated to be less important. The diffraction production of a φ meson in graph (e) with the subsequent decay

$$\varphi \to K_1 + K_2$$

is also estimated to be of lesser importance because of the small partial widths for forming the φ as well as for its subsequent decay*. This leaves the K* meson as the lightest particle with the quantum numbers of the t or momentum transfer channel and we expect K* exchange to be a dominant feature in high energy K^0 photoproduction.

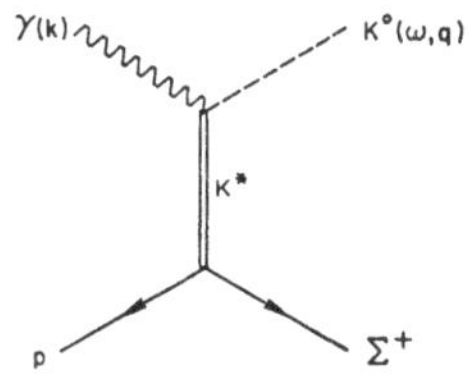

In the spirit of making a prediction of the intensity of the K_2 beam emerging from accelerators such as DESY, CEA, CORNELL, EREVAN, and SLAC the attempt was made to bias calculations to yield a lower limit prediction. In particular only the two body final state in the reaction

$$\gamma + p \to K^0 + \Sigma^+$$

Fig. 7. K* exchange contribution to photoproduction of a K^0 beam in the reaction
$$\gamma + p \to K^0 + \Sigma^+$$

is retained as illustrated in Fig. 7 (for example the amplitude to produce a Y_1^* (1385) is ignored altogether). Also the strong damping of this reaction amplitude due to the effect of many open and competing channels is computed with the distorted wave Born approximation method developed primarily by GOTTFRIED and JACKSON [11].

The applicability of this absorption model to vector exchange processes is not well established at high energies. In the $3-4$ GeV/c range the K* exchange mechanism has had good success [12] in fitting the

_____

* These are known to be $\Gamma_{\varphi \to \pi\rho} < 0.3$ MeV and $\Gamma_{\varphi \to K_1 K_2} < 2$ MeV and so lead to small diffraction production amplitudes in the calculations of Ref. [7].

data on the $p\bar{p} \to \Lambda\bar{\Lambda}$ reaction when the K^* pΛ coupling parameters are specified on the basis of SU_3 symmetry in terms of the ρ nucleon coupling and the maximum possible absorption of the low partial waves (no S wave at all) is introduced in the Gottfried-Jackson model again using SU_3 to relate parameters for the $\Lambda\bar{\Lambda}$ and $p\bar{p}$ channels.

Although this gives us some confidence in the model as applied to photoproduction in the 5 GeV energy range, we are also interested in a much higher range of 10 to 20 GeV with a particular eye on SLAC's hopeful future role as a secondary beam factory. It should be recognized that a model such as this which in effect reduces the Born cross section by an energy independent absorption factor $\eta(b)$

$$\left(\frac{d\sigma}{d\Omega}\right)_{\text{lab}\ \substack{\theta\lesssim 1\,\text{rad} \\ k/M_{K*}\gg 1}} = \frac{1}{2M_{K*}^2}\left(\frac{g_{K*p\Sigma}^2}{4\pi}\right)\left(\frac{g_{K_0^*K_0\gamma}^2}{4\pi}\right)\frac{\theta^2}{[\theta^2 + M_{K*}^2/k^2]^2}\,\eta(b) \tag{1}$$

is inadequate asymptotically [13] as $k \to \infty$. In the approximation of a purely imaginary amplitude for $K^0\Sigma^+$ scattering at high energies s and low momentum transfers t, with the simple parametrization

$$f(s, t) = i\,\text{Im}f(s, t) = i\,e^{-\frac{1}{2}At}\,\frac{s}{8\pi M_\Sigma}\,\sigma_{\text{tot}}(s) \tag{2}$$

the factor $\eta(b)$ is, in the distorted wave Born approximation, for a collision impact parameter $b = 1/|t|^{1/2}$

$$\eta(b) = \left\{1 - \frac{\sigma_{\text{tot}}(s)}{4\pi A(s)}\,e^{-b^2/2A(s)}\right\}. \tag{3}$$

This factor describes absorption of the K^0 wave along its straight line path through the Σ^+ absorbing potential and is energy independent for constant total cross section σ_{tot} and width $A \cong 10$ (BeV^{-2}) of the diffraction pattern. Hence the peak in the differential cross section (1) continues to grow quadratically with increasing energy, the integrated cross section grows logarithmically, and the absorption factor does nothing to tame the difficult high energy limiting behavior of vector exchange amplitudes that motivated the original application of Regge pole ideas at high energies.

A detailed calculation of the K_2 beam has been made [10] with the various helicity amplitudes and individual partial waves in the amplitude for the process of Fig. 7 being projected out and reduced by their appropriate absorption factors. In order to establish a lower limit we set $\frac{\sigma_{\text{tot}}}{4\pi A} = 1$ corresponding to complete S-wave absorption although the experimental parameters predict a value of $\frac{\sigma_{\text{tot}}}{4\pi A} \approx 0.6$.

The $K^{*0} K^0 \gamma$ coupling constant is given in terms of the $\rho \to \pi + \gamma$ radiative decay width, Γ_γ, with the aid of SU_3 symmetry and the assignment of octet transformation properties to the electromagnetic current, by

$$\frac{g_{K*K\gamma}^2}{4\pi} \cong \frac{1}{8}\,\Gamma_\gamma.$$

As a conservative estimate we specify $\Gamma_\gamma \sim 0.15$ MeV and our results may be scaled linearly with the radiative decay width as better direct experimental information is obtained. With this choice the radiative decay width of $K^{0*} \to K^0 + \gamma$ comes out to be ~ 0.2 MeV.

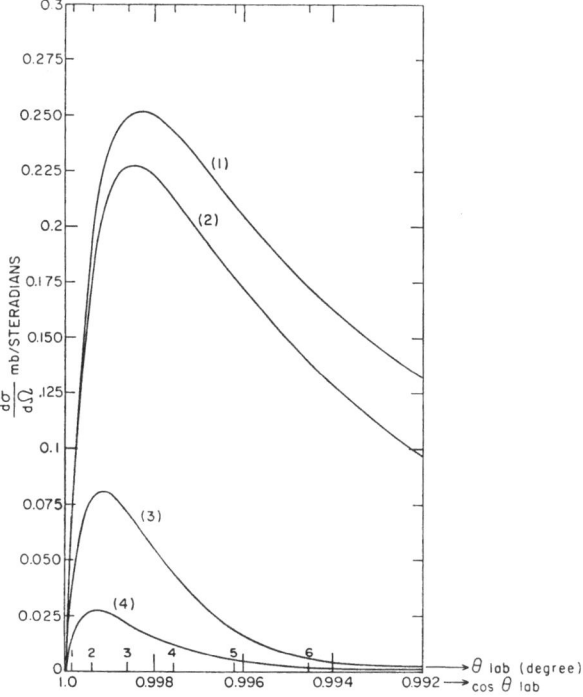

Fig. 8. Laboratory differential cross section at 15 BeV. Curve (1) gives the Born approximation. Curve (2) is obtained after subtraction of the $j = 1/2$ partial wave. Curves (3) and (4) are respectively obtained after the $j = 1/2, 3/2, 5/2, 7/2$, and all partial waves have been corrected for absorption in the final state (curve taken from [10])

Experimental information on Γ_γ will be discussed in the next section. The intensity of K^0's computed in this way is illustrated in the following graph (Fig. 8) for a 15 BeV incident photon and shows the effect of the final state absorption* in reducing the peak beam intensity by a factor

---

* At the peak for K^* exchange, $\theta = M_{K*}/k$ and the impact parameter in Eq. (3) is $b_{max} = M_{K*}^{-1}$ leading to an absorption reduction

$$\eta_{max}(M_{K*}^{-1}) = 1 - e^{-\frac{1}{16}} \sim \frac{1}{16}$$

This has the effect of shifting the peak of the K^0 angular distribution to smaller angles where the differential cross section reaches $\frac{1}{10}$ of the value of the peak in the Born approximation. For one- pion exchange cross sections, the impact parameter at the peak $\theta = \mu/k$ is $b = \mu$ and

$$\eta(\mu^{-1}) \cong 1 - e^{-2} \sim 0.9$$

so that the effect of maximal Gottfried-Jackson absorption on the peak differential cross section is very small for one-pion exchange processes.

of 10. Translating this number to a K^0 beam flux consider $4 \cdot 10^{14}$/sec incident 20 BeV electrons producing photons in a 1/2 radiation length target, as envisaged SLAC operating conditions. If these photons are incident on a 8 grams/cm^2 hydrogen target they will produce 10^6 K_2's/sec $\cdot$ BeV/c at 15 BeV within one millisterad about $\theta \sim 2^0$. This compares favorably to the corresponding flux of K_2^0's at 0^0 expected to be produced by an external proton beam with $5 \cdot 10^{10}$ protons/sec at CERN or Brookhaven [14].

As an independent comment on the factor of 10 damping in our calculation of K_2's due to final state absorption we note that at such high energies the vector exchange processes may be more accurately describable in terms of reggeized trajectories in the spirit of Chew and collaborators. In this case we might be led to modify (1) by a damping factor [15]* of $\left(\dfrac{s}{2M_{K^*}\cdot M_p}\right)^{2[0.54-1.0]} \approx \dfrac{1}{13}$ with very similar numerical results. As final word on the K_2 beam we remark that at the 6 BeV energies of CEA and DESY the peak differential cross section is computed to be ≈ 4 μb/ster, is comparable to the charged $K^\pm$ yields, and should be readily detected.

Of course once we are this far away from the pole, it is a minor additional sin to cast a glance at the nucleon exchange pole and recent results make a few remarks on this topic unmutable. So I ask the forebearance of my theoretical colleagues and remind you that in these considerations it is *only* the very approximate magnitudes of differential cross sections for small values of the momentum transfer that are of interest to us.

Very recently backward peaks have been observed [16] in $\pi^\pm$-proton scattering with cross sections of

$$\left(\frac{d\sigma}{d\Omega}\right)_{180^\circ \text{ cM}} \cong \begin{array}{l} 70 \ \mu b/\text{ster at 4 GeV} \\ 30 \ \mu b/\text{ster at 8 GeV} \end{array}$$

for the π^+-proton case. For π^--proton scattering the backward peaks are smaller by a factor of 5 to 10. If we interpret the π^+ proton scattering in terms of pure one nucleon exchange we find an energy independent result of $\cong 90$ mb/ster which is too large by a factor of one to four thousand. However an application of the Gottfried-Jackson model to the nucleon exchange amplitude has been performed recently by Trefil [17], who has reduced the difference between calculation and observation to less a factor of three and has reproduced approximately the backward angular peaking with the following assumptions:

1) Complete S-wave absorption, corresponding to $\dfrac{\sigma_T}{4\pi A} = 1$ in Eq. (3). This number must be $\gtrsim 0.9$ in order to avoid violating the unitarity limit for S-waves. The reduction in cross section due to the absorption is a factor ≈ 100.

---

* In his report, Prof. van Hove has already noted the similar decrease with p_{lab}^{-1} of cross sections dominated by exchange of vector ρ mesons in the t channel. Such a factor applied to the calculated K_2 beam flux would reduce our estimate of 10^6/sec to $2-3 \times 10^5$/sec and would not alter our primary conclusion.

2) Form factors at the π-nucleon vertex and for the nucleon propagator as computed recently by SELLERI [18] for the nucleon removed by ~ 1 GeV from its mass shell. This virtual nucleon effect reduces the cross section by a factor ≈ 25.

3) Inclusion of N* exchange which adds coherently with N exchange for the π^+ p case and which is the entire story for π^- p scattering. Using similar form factors as for N exchange this adds a contribution of $\sim 40\%$ to the π^+ p process and predicts π^- p cross sections about 1/2 the magnitude as for the π^+ case.

The results of TREFIL are still largely energy independent and so fail to reproduce the observed drop by a factor of $2-3$ in the $4-8$ GeV energy range for π^+ p scattering. This drop-off requires Reggeization of the nucleon trajectory, as for the K* case.

However now that we are within an order of magnitude agreement, speculations on the anti-proton or baryon beam become fair game. The primary differences between the calculations of

$$\pi + p \rightarrow p + \pi$$

and

$$\gamma + p \rightarrow \bar{p} + (n)$$

are that initial state absorption is missing in the photon induced case. Furthermore according to current conservation the charge part of the electromagnetic vertex times the nucleon propagator has no change in value for the nucleon off the mass shell. Retaining however full reduction of the final state amplitude due to the absorption correction as well as the reduction of the form factor at the pp vertex we can say that an anti-nucleon beam should not be reduced by more than a factor of ≈ 200 below the perturbation predictions. This may not be good theoretical physics but is still a tremendous anti-nucleon beam – coming to some few microbarns per ster-BeV at 15 BeV. At SLAC operating conditions, this cross section translates into an anti-proton (baryon) beam flux of $\approx 3 \cdot 10^5$ particles per BeV/c per second into a milli-steradian of solid angle at 15 BeV.

II. Peripheral Processes

We turn next to more detailed analyses of the photoproduction of ρ^0 mesons at 0^0 with the hope of identifying the precise roles of the diffraction and single pion exchange contributions, graphs (b) and (e), in Fig. 1.

It was conjectured that a forward diffraction peak would characterize high energy photoproduction of the zero strangeness neutral vector mesons since they have quantum numbers in common with the photon. The characteristic predictions of this conjecture are:

1) The forward differential cross section $\left(\dfrac{d\sigma}{d\Omega}\right)_{0^0} \propto k^2$, i.e. $\left(\dfrac{d\sigma}{d\Omega}\right)$ increases with the square of the incident photon laboratory energy.

2) $\left(\dfrac{d\sigma}{d\Omega}\right)_{0^\circ} \propto A^{4/3}$ — i. e. the amplitude increases with the nuclear surface area where A is the atomic weight.

3) $\left(\dfrac{d\sigma}{d\Omega}\right)$ decreases exponentially with the square of the momentum transfer, i. e. $\propto e^{-a|t|}$, with a half width ≈ 300 MeV as in (2).

All of these predictions are in approximate agreement with the most recent experiments, both with counters and with the bubble chamber [19] at CEA. These studies have been carried out in the energy range up to 4.4 GeV with H, C, Al, and Cu as target nuclei and all of the above general features of diffraction scattering were found to be reproduced as reported to this conference. The A dependence of the ρ^0 photoproduction is the same as that found for π-nucleus scattering, following along an approximate $A^{4/3}$ curve for C, Al, and Cu. The energy and angular variations are characteristic of a diffraction mechanism as given above.

Moreover the magnitude of the cross sections is very large*; in C the forward differential cross section is

$$\left(\frac{d\sigma}{d\Omega}\right)_{\rho^0,\,0^\circ} = 5 \text{ mb/ster per nucleon at 4.4 GeV}$$

and in H   (4)

$$\left(\frac{d\sigma}{d\Omega}\right)_{\rho^0,\,0^\circ} = 1.3 \text{ mb/ster at 4.4 GeV}$$

with the total elastic cross section in hydrogen integrating to

$$\sigma_{\rho^0} = 25 \ \mu\text{b at 4.4 GeV} .$$

The theoretical conjecture of diffraction production was accompanied [7] by a calculation based on the multiperipheral model of AMATI, FUBINI, and STANGHELLINI [20]. What was computed was the ratio

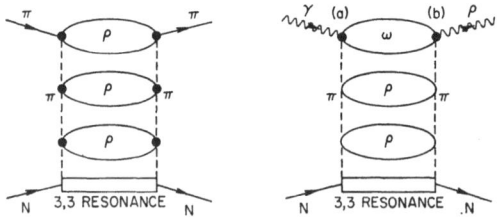

Fig. 9. Graphs for multiperipheral ladders contributing to pion-nucleon diffraction scattering and to diffraction photoproduction of a ρ^0

of the $\gamma + p \to p + \rho^0$ to the $\pi + p \to \pi + p$ diffraction cross section by replacing the top rungs in the multiperipheral ladder as shown in Fig. 9 with the result

$$\left(\frac{d\sigma}{d\Omega}\right)_{\gamma p} = \left(\frac{1 + \cos^2\theta}{2}\right) \beta_\rho \left(\frac{m_\rho}{48\,\Gamma_\rho}\right)^2 \left(\frac{g^2_{\gamma\pi\omega}}{4\pi}\right) \left(\frac{g^2_{\rho\pi\omega}}{4\pi}\right) \left(\frac{d\sigma}{d\Omega}\right)_{\pi N} \qquad (5)$$

---

* These absolute cross sections are taken from the work of PIPKIN and collaborators. (See report of J. J. RUSSELL to Session 1 B.)

where β_ρ, m_ρ, and Γ_ρ are the velocity/c, mass, and total decay width of the ρ^0, respectively; $\left(\dfrac{d\sigma}{d\Omega}\right)_{\pi N}$ is the π-nucleon diffraction cross section at the same energy and angle, $(g^2_{\gamma\pi\omega}/4\pi)$ is the dimensionless coupling constant at the $\gamma\,\pi\,\omega$ vertex and in terms of the $\omega \to \pi^0 + \gamma$ radiative decay width is given by

$$\Gamma_{\omega\to\pi\gamma} = \frac{1}{24}\left(\frac{g^2_{\gamma\pi\omega}}{4\pi}\right)\left(1 - \frac{\mu^2}{m^2_\omega}\right)^3 m_\omega \tag{6}$$

and for

$$\Gamma_{\omega\to\pi\gamma} \sim 1\ \text{MeV}\,,\ \frac{g^2_{\gamma\pi\omega}}{4\pi} \approx \frac{1}{30}\,. \tag{7}$$

The coupling constant at the $\rho\,\pi\,\omega$ vertex $\dfrac{g^2_{\rho\pi\omega}}{4\pi}$ has been estimated [21] from the measured decay width of ≈ 10 MeV for $\omega^0 \to 3\pi$, with the assumption that this decay is dominated by $\omega \to \rho + \pi \to 3\pi$, to be $g^2_{\rho\pi\omega}/4\pi \approx 12$. Alternatively we may use (5) to "measure" this constant by comparison with the experimental result (4). This gives $g^2_{\rho\pi\omega}/4\pi = 10$ in very satisfactory agreement with the above theoretical deduction.

Only in hydrogen does an obvious departure from the diffraction mechanism appear in the form of a slower energy variation of the cross section than with k^2. This is as would be the case if the one pion exchange contribution contributes comparably with the diffraction cross section at the lower energies. These two amplitudes are $90°$ out of phase and so do not interfere. The one pion exchange contribution decreases with increasing energy as

$$\frac{1}{k^2}\,\frac{1}{[1 + (m^2_\rho/2\,k\,u)^2]^2}$$

in the forward direction and for $k > m^2_\rho/2\,\mu = 2$ BeV approaches

$$\left(\frac{d\sigma}{d\Omega}\right)_{ope,\rho^0,0°} = \frac{3}{8}\left(\frac{g^2_{\pi NN}}{4\pi}\right)\left(\frac{\Gamma_{\rho\to\pi\gamma}}{m_\rho}\right)\frac{m^6_\rho}{\mu^4\,M^2_p}\frac{1}{k^2} \tag{8}$$

where $\dfrac{g^2_{\pi NN}}{4\pi} = 15$ is the π-nucleon coupling constant and $\Gamma_{\rho\to\pi\gamma}$ is the ρ radiative decay width, analogous to (6) for the ω. If one assumes that the deviation from the diffraction production can be attributed to pion exchange and fits the observed energy variations in this way a value for $\Gamma_{\rho\to\pi\gamma}$ can be deduced. This analysis has been carried out by PIPKIN et al. [19], with the result $\Gamma_{\rho\to\pi\gamma} = 1.5 \pm 1.5$ MeV.

A similar analysis of ω photoproduction at forward angles leads to the sum of formulas (5) and (8) but with the ω and ρ radiative decay widths interchanged. Measurement of this functional variation and magnitude of the ω photoproduction cross section thus provides a crucial, parameter-free test of this model.

For a number of reasons we have a crucial interest in accurately determining $\Gamma_{\rho\to\pi\gamma}$ by further photoproduction studies and by direct observation of the radiative decay branching ratio [22] or of conversion of a π to a ρ in a Coulomb field [7].

The first of these is that the K_2 beam prediction discussed earlier is based on a "conservatively" assumed value of $\Gamma_{\rho\to\pi\gamma} \sim 0.15$ MeV.

On purely theoretical grounds there are SU_3 and SU_6 estimates that [23]

$$\Gamma_{\rho_0 \to \pi_0 \gamma} = \frac{1}{3}\,\Gamma_{\omega \to \pi_0 \gamma} \sim 0.35 \text{ MeV}$$

or that [24]

$$\Gamma_{\rho_0 \to \pi_0 \gamma} = \frac{1}{9}\,\Gamma_{\omega \to \pi_0 \gamma} \sim 0.1 \text{ MeV}$$

and BRONZAN and LOW [25] have conjectured that the ratio of ρ_0 to ω radiative decay widths will be small on the basis of A parity.

Recently it has been proposed [26] that one can understand the static magnetic moment of the deuteron in the context of a potential model of the deuteron binding and wave function which meets all other known binding and scattering parameters if exchange current corrections to the impulse approximation calculations are included. In particular with the desired D wave probability $P_D = 0.07$ for the deuteron, the moment in the impulse approximation is

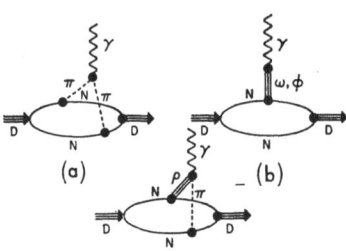

Fig. 10a and b. $\gamma\rho\pi$ exchange current contributing to the deuteron magnetic moment and to elastic electron-deuteron scattering; (a) shows a pion-exchange current that vanishes since the deuteron has isotopic spin zero; (b) shows a contribution to the isoscalar nucleon form factor which is included in the impulse approximation

$$\mu_D = (\mu_{\rm P} + \mu_{\rm N}) - \frac{3}{2} \times \\ \times P_D\left(\mu_{\rm N} + \mu_{\rm N} - \frac{1}{2}\right) \quad (9)$$

and is too small by 2%. To remove this discrepancy within the context of the impulse approximation requires reducing the D wave probability to less than 4%. The alternative solution to this dilemma is provided by a $\gamma\,\rho\,\pi$ exchange current as illustrated in Fig. 10. This is a unique candidate of less than nucleon mass to remedy the moment discrepancy since the deuteron has isotopic spin zero and so only an isotopic scalar current can contribute. Detailed analysis shows that a radiative decay width of $\Gamma_{\rho\pi\gamma} \sim \frac{1}{2} - 1$ MeV supplies the needed addition to the static magnetic moment [26]. New evidence in support of this exchange current contribution was presented to this conference in the analysis by BUCHANAN and YEARIAN of elastic electron deuteron scattering [27]*. They have measured large angle magnetic scattering cross sections that lie more than 50% above the impulse approximation calculations at momentum transfers of $q \sim 3$ (fermi)$^{-1}$. The exchange current which

---

* We emphasize, as stated in Ref. [26], that this exchange current interaction is just one of a number of corrections to the impulse approximation current for Pauli particles that are of the same order in the nucleons' $(v/c)^2$. In including it here we are recognizing that a $\rho \to \pi + \gamma$ radiative decay has a necessary contribution to the magnetic interaction of a deuteron. Its role as a *point* coupling at $q > 0.7\,M = 3f^{-1}$ is much more dubious theoretically and indeed in conflict with data. Professor E. LOMON has just reported that he and Prof. H. FESHBACH at MIT have constructed a deuteron model with just under 5% D-state. According to this model, a somewhat smaller $\Gamma_{\rho\pi\gamma}$ would be needed to explain the static magnetic moment (with a detailed integration of the exchange operator over the wave function necessary before we can say just how much).

supplied the needed static magnetic moment correction of 2% contributes the desired large correction $> 50\%$ at $q \sim 3$ fermi^{-1} because the form factor associated with this contribution is much tighter and does not fall off as rapidly with increasing q since an exchange operator gives heavier weighting to the region of small interparticle separations.

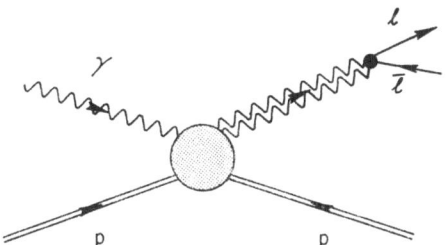

Fig. 11. Virtual Compton terms leading to lepton pair (e$^-$e$^+$ or $\mu^-\mu^+$) production

A final crucial interest in the radiative decay modes of the vector mesons stems from the need to compute their contributions to the large angle pair production cross sections which are being measured as tests of quantum electrodynamics at high momentum transfers. Symmetric $\mu^\pm$ pair and e$^\pm$ pair experiments have been performed [28] at CEA and indications of a deviation from the Bethe-Heitler formula have been observed in the latter case as published by BLUMENTHAL et al. and reported to this conference in Session VI by Professor F. PIPKIN. We may review critically at this point the possible corrections due to strong interactions that might lead to any such deviations via virtual Compton terms of the type illustrated in Fig. 11. Recall that for symmetric pairs these terms do not interfere with the Bethe-Heitler ones and so only ratios of virtual Compton to Bethe-Heitler cross-sections are involved.

First we compute the background due to the total photoproduction cross section neglecting the fact that the virtual photon producing the lepton pair is off its mass shell. Here we can appeal to the Kramers-Kronig relation for forward scattering of light to express the amplitude in terms of the total photo-absorption cross section measured at CEA to be $\lesssim 100\,\mu$b and essentially constant from $1-5$ GeV i. e. we insert ($\alpha = 1/137$)

$$A\left(\omega, 0^0\right) = -\frac{\alpha}{M} + \frac{i\omega}{4\pi}\,\sigma_{\mathrm{tot}}\left(\omega\right) + \frac{\omega^2}{2\pi^2}\,P\int\limits_0^\infty \frac{d\omega'}{\omega'^2 - \omega^2}\,\sigma_{\mathrm{tot}}\left(\omega'\right) . \quad (10)$$

The contribution of the absorptive part dominates and leads to the correction factor, R to the Bethe-Heitler formula for symmetric pairs*

$$R = \left\{1 + \left(\frac{\omega^2\,\sigma_{\mathrm{tot}}}{8\pi\alpha}\right)^2 \tan^4 \theta/2\right\} \quad (11)$$

where ω is the photon energy and the electron and positron, or μ^- and μ^+, emerge each with energy $\frac{1}{2}\,\omega$ and at an angle θ to the left and right of the incident direction. At the highest energies and largest

* This formula is obtained by introducing Eq. (10) in to the calculations in [29].

angles of the electron pair experiments at CEA ($\omega \sim 5.5$ BeV at $\theta \sim 7.5°$) this factor corrects the Bethe-Heitler formula by $2^1/_2\%$. However when one recalls the sharp dip in the cross section at the symmetric point

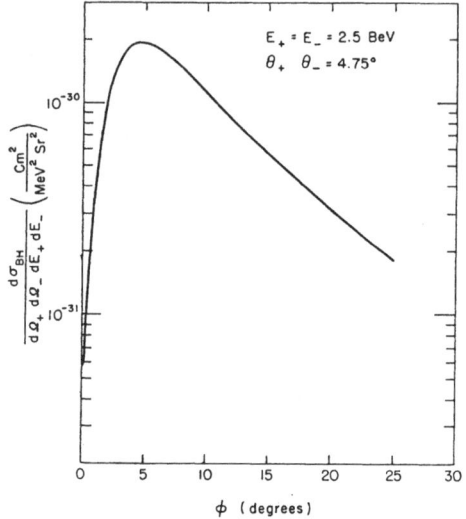

Fig. 12. Example of dip in Bethe-Heitler formula at the symmetric point plotted as a function of angle of azimuthal deviation for equal energies and polar angles. From Harvard Ph. D. thesis of R. Blumenthal. See Ref. [32]

where the transverse current vanishes as illustrated in Fig. 12 and integrates over the experimental resolution which includes much of the bump on either side of the dip, this correction drops effectively to $<0.2\%$. A similar conclusion applies to the μ pair experiment which extended out to $\theta \sim 10°$ with $\omega = 5$ BeV.

We turn then to contributions of the virtual vector mesons and in particular to the ρ^0 which appears to play the predominant role[*]. In terms of the branching ratio for the ρ^0 to decay to an $e^- e^+$ pair [30]

$$\frac{\Gamma_{\rho \to e\bar{e}}}{\Gamma_\rho} \lesssim 10^{-4} \tag{12}$$

we can write the differential cross section for symmetric pairs to be produced via a virtual ρ^0

$$\frac{d^3\sigma}{d\mathscr{E}\,d\Omega_+\,d\Omega_-} = \frac{3}{8\pi^2}\frac{\Gamma_{\rho \to e\bar{e}}}{m_\rho}\frac{k}{m_\rho^2}\frac{1}{(1-s/m_\rho^2)^2 + (\Gamma_\rho/m_\rho)^2}\left(\frac{d\sigma}{d\Omega}\right)_{\rho^0,0^0} \tag{13}$$

where k is the incident photon energy and s is the square of the mass of the pair that is produced; in writing (13) we assume we are near

---

[*] ω^0 photoproduction at forward angles from hydrogen at 4.4 GeV appears to be reduced by a factor of $\approx 1/6 - 1/8$ relative to ρ^0's (F. Pipkin, private communication). If this ratio is transcribed via the diffraction model, Eqs. (5) and (6) to a ratio of radiative decay widths, we find from Eq. (7) that $\Gamma_{\rho \to \pi\gamma} \sim 0.15$ MeV in agreement with the value introduced into the K_2 beam production calculation but smaller by a factor of ~ 3 than that desired in the deuteron-exchange magnetic moment analysis.

the ρ mass, $s \approx m_\rho^2$, and neglect numerator terms $\propto (1 - s/m_\rho^2)$. The value for $\left(\dfrac{d\sigma}{d\Omega}\right)_{\rho^0, 0^0}$ can be taken directly from experiment or expressed via (5) and (6) in terms of the diffraction model. We are interested in the ratio of (13) to the Bethe-Heitler contribution which has the simple approximate form for the symmetric case

$$\frac{d^3\sigma}{d\mathscr{E}\, d\Omega_+\, d\Omega_-} = \frac{\alpha^3}{16\pi^2\, k^3} \frac{\cos^2\theta/2}{\sin^6\theta/2} . \tag{14}$$

For the largest angles and highest energies in the electron pair experiments this ratio reaches to 0.1 but the effective correction is only 1% when the experimental resolution over the dip in the Bethe-Heitler formula is included. Although the observed photo-absorption cross sections fail to provide the explanation for the deviation in large angle $e^-\, e^+$ pairs from quantum electrodynamic predictions, they may begin to play a substantial role at the extreme points of the muon pair experiment which reaches to a higher value of the mass of the pair $s \sim m_\rho^2$ and, depending on the experimental resolutions, contribute significantly[*].

In concluding this section we call attention to a particular interest in detecting and studying the charged pion production cross sections at the precisely forward angle $\theta = 0°$. The peripheral production processes all lead to peaks at $\theta \sim m/\omega_q$, where ω_q is the energy of the high energy pion emerging at angle θ and m is the mass of the particle exchanged in the t channel with the target nucleons, and within this angle the cross section dips toward zero [32], decreasing as θ^2. This forward decrease results from the impossibility of conserving the spin angular momentum of the incident transverse photon for a spin zero pion emerging at $0°$ unless the amplitude is accompanied by nucleon spin flip. For single particle exchanges this spin flip occurs with an amplitude proportional to the momentum transferred to the nucleon, which has a minimum value in the forward direction of

$$|t_{\mathrm{Min}}|^{1/2} \approx \mu^2\, \frac{\Delta E}{k} + (\mu^2/2k)^2 \ll \mu^2$$

where k is the incident photon energy and $\Delta E = k - \omega$ the energy transferred to the nucleon in producing the pion with mass μ and energy ω, if vector or spin zero mesons are exchanged in the t channel. Non peripheral processes escape this suppression since there are large matrix elements for spin flip into and/or out of high energy intermediate nucleon states as in Fig. 13. This is also true of peripheral production of pseudo-scalar mesons via an axial vector exchange since the nucleon matrix element then has the familiar Gamow-Teller form of β-decay and the amplitude may be written $\langle\vec{\sigma}\rangle \cdot \vec{\varepsilon}_\gamma$. The very forward angles of pion (or K meson) photoproduction at high energies thus are very interesting,

---

* KRASS [31] has made some detailed calculations of possible corrections but with more emphasis on the one-pion exchange contributions as the large diffraction production was not known at the time of this work. The large vector meson photo-production cross sections also may mean sizable contributions to the very asymmetric μ pairs [S. D. DRELL, Phys. Rev. Letters 13, 257 (1964)], and this point is presently under further study (R. PARSONS).

though difficult, for experimental study since they probe the non-peripheral features of the production amplitude as well as a possible exchange of axial mesons which we are tempted to associate with the structure of

Fig. 13. Non-peripheral contribution to photo-pion production

the axial current in the same way as the ρ, ω, and φ enter the vector current structure and knowledge of both energy variations and branching ratios will be of great value.

III. Central Collisions

We also avoid the peripheral plateau by looking at large angle or momentum transfer events. In these central collisions it is possible to test symmetry schemes without major corrections due to mass splittings between different particles assigned to the same multiplets; i. e. under conditions such that $|t|$ and $s \gg \Delta M^2$. Such tests have been proposed by LEVINSON, LIPKIN, and MESHKOV [34] for the octet model in the SU_3 symmetry group. In meson-nucleon scattering these include the equalities

$$d\sigma(K^- p \to \pi^+ \Sigma^-) = d\sigma(K^- p \to K^0 \Xi^0)$$
$$d\sigma(\pi^- p \to K^+ \Sigma^-) = d\sigma(K^- N \to K^0 \Xi^-) \qquad (15)$$
$$d\sigma(\pi^- p \to \pi^+ N^*_{3/2} (1238)) = 3 d\sigma(\pi^- p \to K^+ Y^{*-}_1 (1385)) \,.$$

In general, simple equalities such as (15) do not emerge from the unitary symmetry model alone since there are a number of open channels through which the reaction can proceed and their relative phases and magnitudes require the input of dynamical assumptions. Formally, this is stated in the observation that both meson and baryon form octet representations in SU_3 and their product can form 1, 8, 8′, 10, $\overline{10}$, and 27 dimensional representations. The reaction can thus proceed through any of six channels and their relative amplitudes together with five relative phase factors at any energy determine the branching ratios. Therefore, analyses of these two-body reactions have heretofore contributed little to our confidence in SU_3 which derives largely from the great success in classification of multiplets and in predicting mass splittings within the individual multiplets. Moreover, the intensity of incident meson beams at high energies has been limited so that only a negligible number of events are observed in the laboratory under the condition of large t as desired to avoid large distortions due to mass splittings and kinematic factors from the exact SU_3 as a symmetry in high-energy scattering processes.

Turning to photon initiated reactions $\gamma +$ Nucleon $\to$ Meson $+$ $+$ Baryon we call attention to the important practical fact that a very intense current of electrons, and in particular of 20 BeV electrons as anticipated at SLAC when operative, leads to a photon flux of sufficiently high intensity to more than compensate for the appearance of a fine structure constant $\alpha = 1/137$ in the ratio of the photon to meson cross sections. We must first determine the transformation properties of the electromagnetic current in the unitary symmetry scheme. We can then analyze whether two body reactions of this type can provide useful information or checks on the symmetry scheme.

In Lagrangian models of the SU_3 symmetry scheme for elementary particles, it is most natural to introduce the electromagnetic current as a unitary octet. However, it is also possible for the electromagnetic current to have a unitary-singlet component and independent evidence on the transformation properties of the current is desired. The following relations between magnetic moments and between transition amplitudes have been proposed [35] as tests of the assumption that the electromagnetic current is a pure octet and a U-spin scalar

$$\mu_N = 2\mu_\Lambda$$
$$\langle \rho^0 \,|\eta\, \gamma\rangle = \sqrt{3}\, \langle \rho^+ \,|\pi^+\, \gamma\rangle . \tag{16}$$

In calculating matrix elements, this is equivalent to equating a photon to the neutral member of the isotopic triplet, ρ^0, and to the isotopic singlet, φ, in the vector meson octet according to the relation.

$$|\gamma\rangle = \left\{ |\rho^0\rangle + \frac{1}{\sqrt{3}}\, |\varphi\rangle \right\} . \tag{17}$$

I would now like to point out that if the role of SU_3 is quantitatively supported by reactions (15) at high energy and the identification of the photon as in (17) established by processes (16) we can propose a feasible program for testing the validity of the statistical model in high-energy collisions*. The idea presented here is to check the very general premise of the statistical model that all open channels should contribute with equal probabilities and with random relative phases, independent of more detailed dynamical questions of specific energy or angle variation of the cross sections.

It is in the central collisions with large s and t that we anticipate the possibility that the concepts of the statistical model may find their natural application. As in the low-energy nuclear physics domain (aside from the direct interaction processes) the colliding particles may be envisioned as forming a compound system with many channels leading to the various possible final state configurations. Aside from phase space and other kinematic factors, the various open reaction channels should be excited with equal probabilities and random relative phases in a statistical model.

---

* This discussion parallels that in a forthcoming paper that discusses both photon and meson initiated reactions according to the statistical model by S. D. DRELL, D. SPEISER and J. WEYERS.

This is the very basic general assumption underlying a statistical model. The statistical model has other characteristic predictions with regard to energy and angle variations of elastic cross sections and of multiplicities, in addition, for inelastic ones. These features, however, are tied to various models and "plausible" dynamical assumptions. Recently arguments have been put forward by BETHE and by WOO [36] pointing out the difficulty of reconciling the observed precipitous drop with energy of the large angle component of the elastic cross section with the statistical model. It is at present not at all clear whether or not the experimental data should be interpreted as indicating the presence of a statistical component in high-energy collisions.

The two-body reactions involving incident meson or photon beams

$$\text{Meson} + \text{Nucleon} \to \text{Meson} + \text{Baryon (resonance)}$$

$$\gamma + \text{Nucleon} \to \text{Meson} + \text{Baryon (resonance)}$$

can proceed through many channels with different quantum numbers. It is the relative roles of channels of different angular momenta that control the angular and energy behaviour of cross sections, and of channels with different internal symmetry quantum numbers that determine the branching ratios for the production of final baryons and mesons with different charge or hypercharge quantum numbers. It is upon these branching ratios that we wish to focus attention. When these cross sections at large s and t are averaged over energy and momentum transfer intervals large compared with the mass splittings within the individual multiplets ($\Delta t^{\frac{1}{2}}$, $\Delta s^{\frac{1}{2}} > \Delta M$), the branching ratios should be determined solely by the combination coefficients, i. e., the appropriate Clebsch-Gordan coefficients — to form the different SU_3 channels. That each SU_3 channel contributes with equal amplitude and random phase is the very basic and sole feature of the statistical model on which we base our predictions.

Some of the practical results of these considerations for photons are summarized in the following two tables. The parameter α appears as a mixing parameter for the two independent channels of a meson-nucleon system that transform as an octet. Denoting the corresponding states by $|8\rangle$, to which we assign the meson and nucleon octets, and $|8'\rangle$, respectively, we form the linear combinations

$$8_1\rangle = \cos\alpha \, |8\rangle - \sin\alpha \, |8'\rangle$$

$$8_2\rangle = \sin\alpha \, |8\rangle + \cos\alpha \, |8'\rangle \, .$$

The rotation angle α is defined by the condition of orthogonality

$$\langle 8_1 | 8_2 \rangle = 0$$

and $\alpha = 0$ if the additional symmetry of R invariance [37] is invoked. Whereas the consequences of R symmetry are unwelcome at low energies it is possible that R may emerge as an approximate symmetry operation at high energies. If the special relations between cross sections that are independent of α are verified by experiment and confirm the role of the statistical assumption in high energy central collisions, it will be possible to determine α from the general ratios.

Table 1

$\gamma P \to \pi^+ N$	$34/225 + 19/225$	$\sin^2 2\alpha - 2\sqrt{5}/225$	$\sin 4\alpha$
$\to \pi^0 P$	$139/450 + 19/450$	$\sin^2 2\alpha - \sqrt{5}/225$	$\sin 4\alpha$
$\to K^+ \Sigma^0$	$139/450 - 11/450$	$\sin^2 2\alpha + 2\sqrt{5}/225$	$\sin 4\alpha$
$\to K^0 \Sigma^+$	$34/225 - 11/225$	$\sin^2 2\alpha + 4\sqrt{5}/225$	$\sin 4\alpha$
$\to K^+ \Lambda$	$31/150 + 1/150$	$\sin^2 2\alpha - \sqrt{5}/75$	$\sin 4\alpha$
$\to \eta P$	$31/150 - 3/50$	$\sin^2 2\alpha$	

$$\frac{2[\gamma P \to K^+ \Sigma^0] - [\gamma P \to K^0 \Sigma^+]}{2[\gamma P \to \pi^0 P] - [\gamma P \to \pi^+ N]} = 1$$

$$\frac{3[\gamma P \to \pi^0 P] - [\gamma P \to K^+ \Lambda] + 2[\gamma P \to \eta P]}{2[\gamma P \to \pi^0 P] - [\gamma P \to \pi^+ N]} = 17/7$$

Table 2

$\gamma P \to \pi^+ N^{*0}_{3/2}$	$7/50 + 4/225$	$\sin^2 2\alpha + \sqrt{5}/225$	$\sin 4\alpha$
$\to \pi^0 N^{*+}_{3/2}$	$41/1200 + 8/225$	$\sin^2 2\alpha + 2\sqrt{5}/225$	$\sin 4\alpha$
$\to \pi^- N^{*++}_{3/2}$	$49/200 + 4/75$	$\sin^2 2\alpha + \sqrt{5}/75$	$\sin 4\alpha$
$\to K^+ Y^{*0}_1$	$7/100 + 2/225$	$\sin^2 2\alpha + \sqrt{5}/450$	$\sin 4\alpha$
$\to K^0 Y^{*+}_1$	$169/600 + 4/225$	$\sin^2 2\alpha + \sqrt{5}/225$	$\sin 4\alpha$
$\to \eta N^{*+}_{3/2}$	$11/48$		

$$\frac{2[\gamma P \to \pi^+ N^{*0}_{3/2}] - [\gamma P \to \pi^0 N^{*+}_{3/2}]}{[\gamma P \to K^0 Y^{*+}_1] - 2[\gamma P \to K^+ Y^{*0}_1]} = 59/34$$

$$\frac{[\gamma P \to \pi^+ N^{*0}_{3/2}] + [\gamma P \to \pi^0 N^{*+}_{3/2}] - [\gamma P \to \pi^- N^{*++}_{3/2}]}{[\gamma P \to \pi^0 N^{*+}_{3/2}] - 2[\gamma P \to \pi^+ N^{*0}_{3/2}]} = 17/59$$

Note added in proofs:

The pion flux from the 2π decay of diffraction produced ρ^0's is calculated to be the dominant source of the observed pions in the experiment of Ref.[2] by NANCY HICKS and Professor F. PIPKIN. In the one-pion exchange model flux of pions at the peak angle of $\theta = m_\pi/k$ exceeds the diffraction produced ones at their peak angle of

$$\theta = \frac{m_\rho}{k} \sqrt{\frac{k - \omega_-}{k}}$$

by a factor of ≈ 2; in Be and heavier targets the coherent amplitude for pion production which increases $\approx A$ becomes dominant.

References

[1] DRELL, S. D.: Phys. Rev. Letters 5, 278 (1960)
[2] BLUMENTHAL, R., et al.: Phys. Rev. Letters 11, 496 (1963)
    BLANPIED, W., et al.: Phys. Rev. Letters 11, 477 (1963) and report to this conference by V. W. HUGHES
    KILNER, J., R. DIEBOLD, and R. L. WALKER: Phys. Rev. Letters 5, 518 (1960)
[3] BALLAM, J. S.: 1960 Summer Study (M-200), Hansen Labs of Physics, Stanford University
    DRELL, S. D.: Rev. Mod. Phys. 33, 458 (1961)

[4] ITABASHI, K.: Phys. Rev. **123**, 2157 (1961)
HADJIOANNOU, F.: CERN Report 1962, and Ph. D. Thesis, Physics Department, Stanford University (1961) (quoted in [3])
THIEBAUX JR., M.: Phys. Rev. Letters **13**, 29 (1964)
LAKERASHVILI, L., and S. MITINYAN: Soviet Physics JETP **14**, 195 (1962)
For nuclear targets see also J. S. BELL: Phys. Rev. Letters **13**, No. 2 (1964)
[5] STICHEL, P., and M. SCHOLZ: Nuovo cimento **34**, 138 (1964)
[6] JONES, L.: Phys. Rev. Letters **14**, 186 (1965)
[7] BERMAN, S. M., and S. D. DRELL: Phys. Rev. **133**, B 791 (1964)
[8] BARGER, V., and M. EBEL: Phys. Rev. **138**, B 1148 (1965)
MAOR, U.: Phys. Rev. **135**, B 1205 (1964)
[9] GREENBERG, J. S., et al.: Phys. Rev. Letters **14**, 741 (1965), and private communications by V. W. HUGHES
[10] DRELL, S. D., and M. JACOB: Phys. Rev. **138**, B 1312 (1965)
[11] GOTTFRIED, K., and J. JACKSON: Nuovo cimento **34**, 735 (1964)
[12] COHEN, G., and H. NAVELET: Saclay preprint (1964) (to be published)
[13] See also the discussion of HOGAASEN, H., and J. HOGAASEN, Nuovo cim. **39**, 941 (1965)
[14] DEKKERS, D., et al.: Report to the Sienna Conference (1963), CRONIN, J. W.: BNL 7957 (May 1964)
[15] PHILLIPS, R. J. N., and W. RARITA: Phys. Rev. **139**, B 1336 (1965)
[16] Private communication, W. KRISCH to M. PERL. See invited talk of L. VAN HOVE
[17] TREFIL, J.: Work in progress at Stanford
[18] SELLERI, F.: Cornell preprint
[19] PIPKIN, F.: Private communication, and J. RUSSELL: Report to this conference
ROGERS JR., A. H.: MIT Physics Department thesis (1965)
[20] AMATI, D., S. FUBINI, and A. STANGHELLINI: Nuovo cimento **5**, 896 (1962)
[21] DASHEN, R. F., and D. H. SHARP: Phys. Rev. **133**, B 1585 (1964)
[22] FIDECARO, G.: Private communication (in progress)
[23] OKUBO, S.: Phys. Letters **4**, 14 (1963)
[24] BADIER, S., and C. BOUCHIAT: ORSAY preprint (1965)
THIRRING, W. E.: Vienna preprint
[25] BRONZAN, J., and F. LOW: Phys. Rev. Letters **12**, 522 (1964)
[26] ADLER, R. J., and S. D. DRELL: Phys. Rev. Letters **13**, 349 (1964)
— Ph. D. thesis, Stanford Physics Department 1964 and SLAC report
[27] YEARIAN, M.: Report to Session 1 A
[28] DE PAGTER, J., et al.: Phys. Rev. Letters **12**, 739 (1964)
BLUMENTHAL, R., et al.: Phys. Rev. Letters **14**, 660 (1965)
[29] BJORKEN, J., S. DRELL, and S. FRAUTSCHI: Phys. Rev. **112**, 1409 (1958)
[30] ZDANIS, R. A., L. MADANSKY, R. W. KRAEMER, and S. HERTZBACH: Phys. Rev. Letters **14**, 721 (1965)
[31] KRASS, A.: Phys. Rev. **138**, B 1268 (1965)
[32] MAOR, U.: Phys. Rev. **135**, B 1205 (1964)
[33] See invited paper by L. OSBORNE to this conference, p. 91
[34] LEVINSON, C. A., H. J. LIPKIN, and S. MESHKOV: Phys. Letters **1**, 44 (1962); Phys. Rev. Letters **10**, 361 (1963)
[35] OKUBO, S.: Phys. Letters **4**, 14 (1963)
[36] BETHE, H. A.: Nuovo cimento **33**, 1167 (1964)
WOO, C. H.: Phys. Rev. **137**, B 449 (1965)
[37] See for example GELL-MANN, M., and Y. NE'EMAN: The Eightfold Way. New York: Benjamin and Co. 1964

Prof. Dr. S. D. DRELL
Stanford Linear Accelerator Center
Stanford University
Stanford/California

Photoproduction of Mesons in the GeV Range

L. S. Osborne

Laboratory for Nuclear Science
Massachusetts Institute of Technology
Cambridge, Massachusetts

I. Introduction

Photoproduction of strongly interacting particles involves the complications of strong interactions with an added electromagnetic vertex; its complete understanding implies a complete knowledge of strong interaction dynamics [1]. This paper will attempt a brief review of the still small amount of experimental material and confront it with various models; in the interests of economy, it will only touch upon some material already presented to the conference.

The main theoretical and empirical models are

(1) the O. P. E. model (more generally, the one-meson-exchange model), which was developed by S. DRELL [2] for high-energy multichannel processes;

(2) resonance production and decay;

(3) Regge pole analysis of photoproduction cross-sections;

(4) symmetry schemes.

II. Drell Processes

Some of the first experiments [3, 4, 7] on π and K photoproduction were done specifically to compare Drell's calculation with experiment, with the added interest that such copious production would have practical consequences at electron accelerators anticipating meson beams. A sample of experimental results is shown in Figs. 1 [3] and 2 [7]. As theories go in high-energy physics, rough agreement between Drell's calculations and these data indicates that the postulated mechanism is in fact operating.

However, several cautionary remarks are in order. Other production mechanisms [3] also exist, such as the large amount of ρ^0 production [5, 6]. In the limit that π^+ and π^- cross-sections become equal (as well as K^+ and K^-), the Drell mechanism predicts that π^+ and π^- photoproduction yields become equal. This is clearly not true for K^+ and K^-

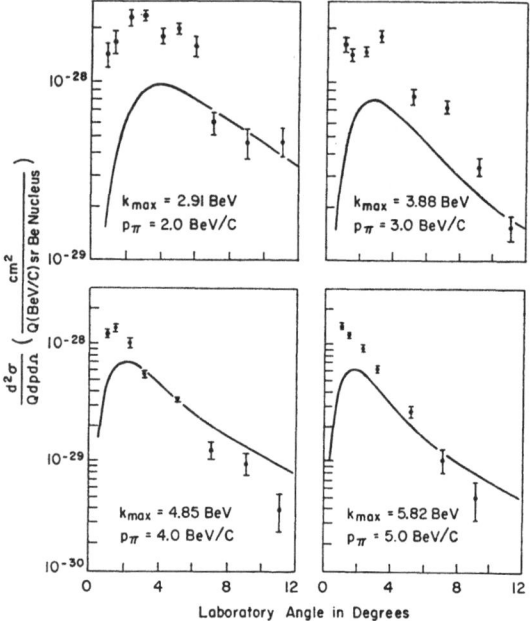

Fig. 1. π^- photoproduction yields from a beryllium target (ref. [3])

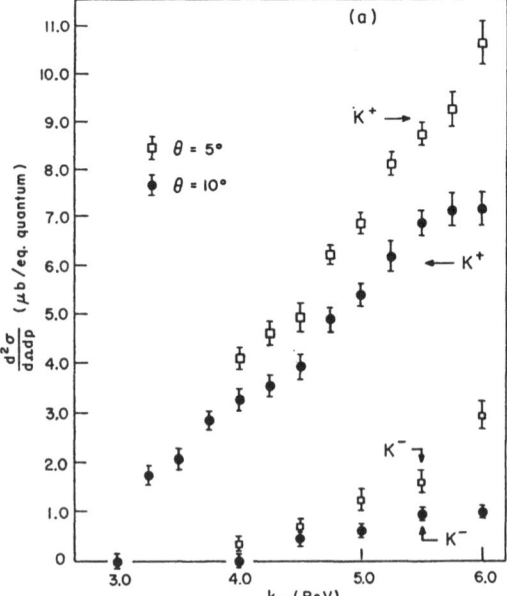

Fig. 2. K^+ and K^- photoproduction yields from hydrogen (ref. [7])

yields but is easily understandable in terms of the fewer channels available for K⁻ production at this energy. This, in turn, is a reflection of an energy too low to apply the Drell model with confidence. If one accepts the interpretation proposed by the Yale group [7] for their K⁺ yields in terms of $K^+ - Y^*$ (new hyperon states) production, then the predominance of such channels in photoproduction, though unseen in the K⁻ − p total cross-section, is in tacit disagreement with the Drell process. In summary, a Drell process seems to appear in the π photoproduction, but other mechanisms are probably present.

III. Resonance Production

Photoproduction and scattering of π mesons below 1 GeV is dominated by the three π-nucleon resonances in that region, and they determine the greater part of the structure of the cross-sections. Above 1 GeV it appears that other mechanisms become more important: the most dramatic evidence for this comes from π-nucleon total cross-sections where such bumps as exist [10, 12] represent a much reduced fraction of the total cross-section. Nevertheless, in the energy region up to 6 GeV, new resonances have been found, and an adequate description of cross-sections must take them into account. An illustration of the structured character of cross-sections is given by Fig. 3, which is a representation of $d\sigma/d\Omega$ at 90° (center -of-mass) for π^0 photoproduction from threshold to 5 GeV. It is not a simple smooth cross-section; most of the structure can be ascribed to known resonances. The question remains whether such behavior continues at higher energies.

A further study of resonances in photoproduction is illustrated by data from the Cambridge Bubble Chamber Group [5]. They consider that their data on $N^{*++} - \pi^-$ production are compatible with production through the excitation of $N^*_{1/2}$ (1512) and $N^*_{1/2}$ (1688); this provides further justification for the description of a large part of photoproduction processes below 1.5 GeV in terms of simple two-body channels where the outgoing "particles" may be unstable states.

At higher bombarding energies, data on resonance excitation by photons comes from our group at M. I. T. [8]. This group found the first evidence of a nucleon resonance of mass approximately 2.75 GeV [9]; further evidence was found subsequently in the π^+ p total cross-section [10]. Our data consisted of differential cross-sections for π^0 photoproduction taken at 60° and 90°, center-of-mass, and is shown in Fig. 4. The 2.75 GeV resonance location is marked with an "f" on this figure.

Also shown in Fig. 4 is some new data taken more recently; it includes 90°, π^0 cross-sections, taken above 2.3 GeV and cross-sections between 3.0 and 3.7 GeV at 120°. This more recent data was taken with a 48 lead-glass-counter hodoscope [11] which enabled us to detect sometimes both gamma-rays and sometimes one gamma-ray from the produced π^0; in both cases a check can be made that the event corresponds to single π^0 photoproduction. The cross-sections computed either from

1-photon or 2-photon detection are in agreement; the cross-sections are generally lower than those of Reference [9], particularly at the minimum at 3.4 GeV. We presume that some contamination from other than single π^0 photoproduction was present in the earlier results; the gamma detector used there was a single total absorption counter. The remainder of the apparatus, the magnetic spectrometer and associated electronics, is

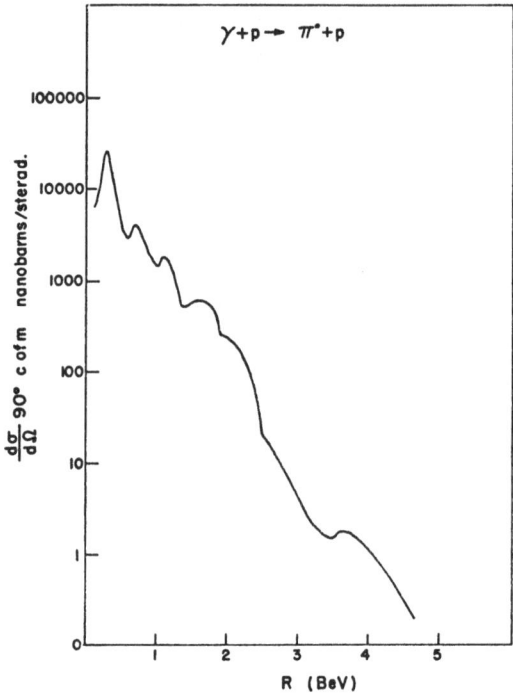

Fig. 3. A simplified sketch of $d\sigma/d\Omega$ (90°) for $\gamma + p \rightarrow \pi^0 + p$. R = Incident Photon Energy

described briefly in Reference [9]. Acquisition of the new data was greatly facilitated by an on-line computer, the Linc II [13], which gave a running analysis of the data as it was being acquired.

The locations of known resonances are labeled by the letters a through f in Fig. 4. There appear to be bumps at 90° corresponding to $N_{1/2}^*$ (1688), (a), and $N_{3/2}^*$ (1920), (b); however, the $N_{1/2}^*$ (2190), (c), and the $N_{3/2}^*$ (2360), (d) appear, if at all, as a single broad bump at 60°. These two resonances also appear unresolved in the $\pi^- p$ total cross-section [12]. There is no particular evidence for the $N_{1/2}^*$ (2700) [10]. In our previous publication we referred to the bump in the 60° data at 2.9 GeV as a possible new resonance; another explanation for such a bump will be mentioned below.

Because of interference effects, a bump in a differential cross-section, probably caused by the interference between a resonant and non-resonant term, may peak at a different energy than the one measured in

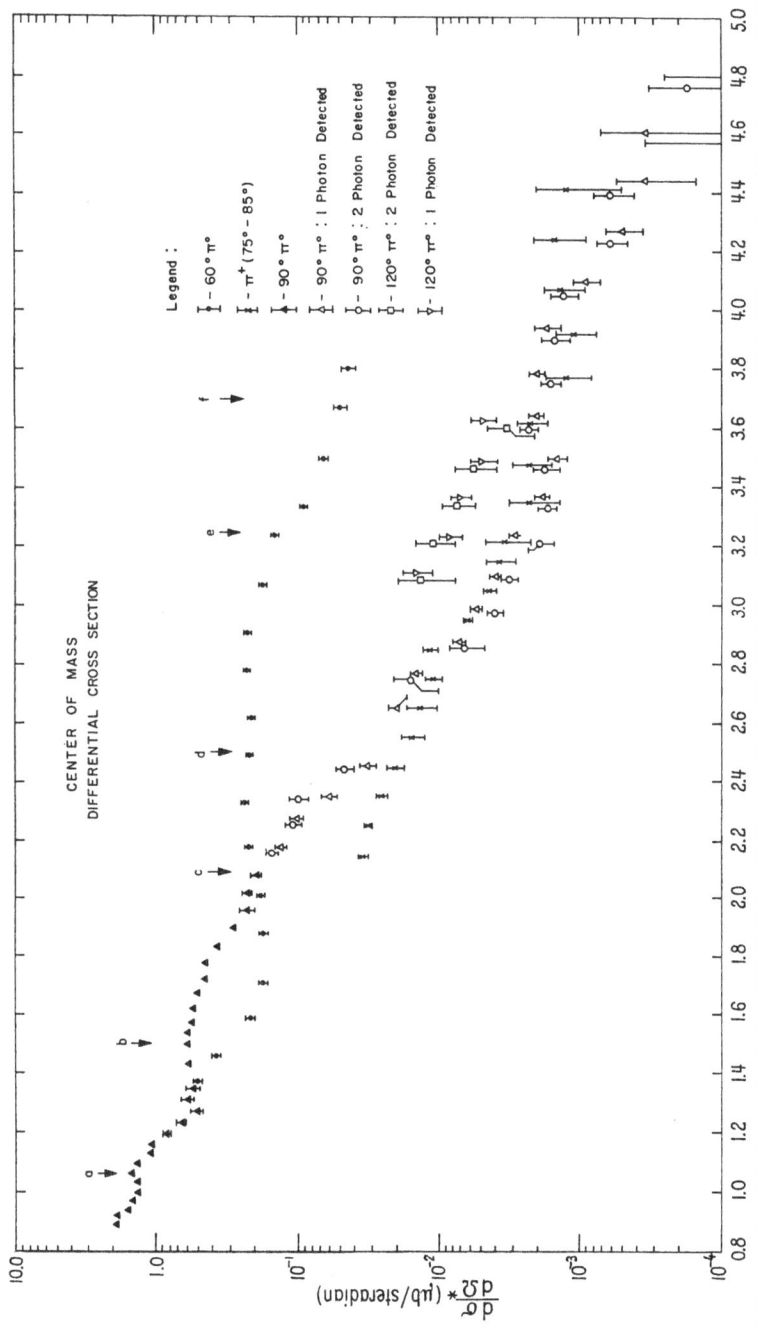

Fig. 4. Data on π^+ and π^0 photoproduction from M.I.T. photoproduction group. Part of the data is from ref. [9]; the other is unpublished. Energy scale: γ-laboratory energy in GeV

total π p cross-section. Höhler [14] has analyzed the behavior of the real and imaginary parts of π p scattering. If one assumes that the main interfering amplitude is real and the π p scattering bump represents the peaking of the imaginary part, then the photoproduction cross-section peak might be displaced as much as a resonance half-width away. Thus Höhler suggests that bumps appearing a few hundred MeV away may be evidence of this effect.

Fig. 4 also shows more recent results for π^+ photoproduction. This data was obtained as a by-product of the π^0 experiment; the π^+ is detected in the spectrometer simultaneously with the protons from π^0 photoproduction. Thus the center-of-mass production angle changes with energy from 75° at 2.1 GeV to 85° at 4.4 GeV. Outside of the monotonous fall with energy, there may be some suggestion of the $N^*_{3/2}$ (2750) at a photon energy of 3.6 GeV.

In Reference [9], we found a bump in the yield of protons in coincidence with low-energy photons at a bombarding energy of 3.6 GeV; this was interpreted as the possible production of η^0. To yield a bump from a continuous bombarding spectrum, the production cross-section must also be going through a peak. Assuming η^0's were the cause, we obtained a cross-section of approximately 80 nanobarns. More recently, we returned to measurement of η^0's using the new gamma-ray detector to reveal the η^0 through the two-gamma decay. We found no evidence for η^0 production and estimate its production cross-section at less than 3 nanobarns. This eliminates the original interpretation that the resonance at 3.7 GeV might be a member of an $SU(3)$, 27 group, $T = 1/2$, resonance whose decay branching ratio η^0/π^0 is 27 to 1. We could not reproduce the original bump in proton gamma coincidences even though the new gamma counter could roughly simulate the previous detecting geometry.

IV. Regge Analysis

It appears that single meson photoproduction at high energies exhibits the same forward peaking as is evident in production by meson bombardment. For the ρ meson this has been seen by the Cambridge and DESY bubble chamber groups [5]. The π^0 photoproduction shown in Fig. 4 also shows a large drop in the cross-section from 60° to 90° at high energies. This general behavior is not unexpected from reasonable models which follow the rule that nature abhors a high-momentum transfer. A simple field-theoretic calculation involving meson production with vector particle exchange gives cross-sections that are much too high. Photoproduction data even as low as 1.2 GeV, for example, gives a faster fall-off with momentum transfer than calculated [15]. For meson-induced reactions this behavior has been variously explained by Reggiazation [1], form factors [16], and absorption models [17]. In the case of ρ photoproduction both the absorption [18] model and a diffraction model [19] have been compared to the data [5] with results somewhat favoring the latter.

For the purposes of this paper we adopt the Regge formalization not so much because it is necessarily correct but because it offers a handy parameterization and offers a link to the language of particle exchange which we assume to be correct in some sense. Furthermore, the bombarding energies we speak of (1 to 5 GeV) are somewhat low to be able to utilize a Regge approach whose success presumes measurements in the limit of very high energies and small angles. Theoretical work has been done on the Regge approach to photoproduction [20–22].

Table 1. *Possible exchanged particles for various processes*

Incoming-outgoing particle	$\pi^+ \pi^+$ / $\pi^- \pi^-$	$\pi^- \pi^0$	$\pi \rho$	$\gamma \pi^+$	$\gamma \pi^0$	$\gamma \rho$	$\gamma \omega^0$
Exchanged boson	P, ρ	ρ	π, ω^0	ρ^-, π^-	ρ^0, ω^0	π, P	π^0, P
Exchanged baryon	N*	N*	N*	N*	N*	N*	N*

P — Pomeranchukon, N* — nucleon or nucleon isobar

Table 1 summarizes the simplest particle exchanges for various reactions; it includes the more common candidates but not the higher-mass mesons. The particle label refers to a trajectory; thus the f^0 is presumably included in P. We have not excluded vertices that violate the Bronzan and Low A quantum number [23]. Indeed, we prefer to evaluate the experimental information as a test of its validity. We note, then, that there are no two processes with the same particles exchanged except for $\gamma \rho$ and $\gamma \omega^0$. Unfortunately the data on ω^0 photoproduction is still meager [5]; one could ascribe its lower production (a factor of about 5) to the operation of the A quantum number. On the other hand, it might be a reflection of reasonably smaller coupling constants.

Further evidence can be obtained on the important trajectories by comparing π^+ and π^0 photoproduction. Looking at Fig. 4, the π^+ photoproduction cross-section is lower than or equal to the π^0 cross-section. In particular, in the region around 2.1 GeV it is approximately a factor of 5 below both 60° and 90° π^0 cross-sections. Such behavior is most easily explained if the $\gamma \rho \pi$ vertex is weak compared to the $\gamma \omega \pi$, which is a further indication that A quantum-number conservation is playing a role. The closer equality of π^+ and π^0 cross-sections at 3 to 4 GeV would come about from the resonance-decay branching ratios required by isotopic spin.

To pursue a different direction we have arbitrarily compared the photoproduction cross-sections of Fig. 4 with corresponding values of π-scattering cross-sections where such data is available (Fig. 5). Though Table 1 suggests no parallel mechanism between the two, it is amusing to note that the ratio, π-scattering $/\gamma \pi$, is approximately $2\pi/\alpha$ even though the cross-sections themselves vary over a factor of 10^3. It is not unreasonable to suppose that the strong-interaction mechanisms present in π-scattering are manifest in photoproduction. One can look at it another way; the inverse process, radiative π capture, is a small inelastic channel, approximately $2\pi/\alpha$ lower than the π-nucleon scattering channel. This leads us to look more closely at π-scattering experiments, many of which are plotted in Fig. 6.

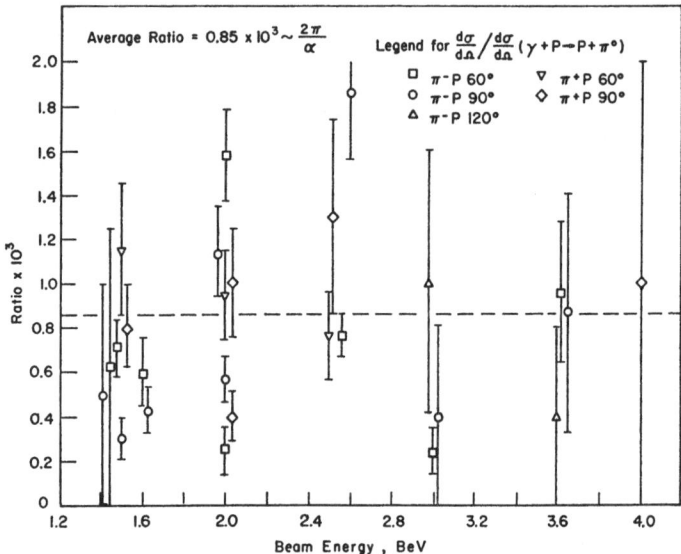

Fig. 5. Ratio of π N scattering cross-sections to π⁰ photoproduction cross-sections

In order to avoid confusion only one representative point from each experiment is shown. In exchange for this clarity a certain subjectivity is introduced in describing the angular distributions with a line. The reader is asked to tolerate this in order to get the general point that the angular distributions are not necessarily smooth curves but frequently show a dip (at $-t \approx 0.6$ GeV/c^2) and then a second maximum $(-t \approx 1.2$ GeV/c^2).

The latter had been seen at 2.08 GeV [24] and explained as an effect due to the $N^*_{3/2}$ (1920) resonance; on the other hand SIMMONS [25] noted that a diffraction model would also suffice to explain it. In favor of the latter argument is the persistence of the dip at the same momentum transfer in various experiments, for example, π charge exchange (see Fig. 6). Such a "diffraction" dip is also seen in other high-energy experiments [26]. On the other hand a second glance at Fig. 6 reveals that this structure may not be present consistently. In any event one must be prepared to find some such structure in photoproduction experiments. Such structure might then explain the rise in the cross-section for π⁰ photoproduction at $\theta = 60°$, $k = 2.9$ GeV. The data of Fig. 4 are plotted in Fig. 7 versus $-t$. We see this bump occurring at $-t = 1.2$ GeV/c^2 where one finds the secondary bump in π scattering. Further measurements will have to be made to ascertain the presence and origin of this structure. Such structure is not simply amenable to an explanation in terms of Regge trajectories; however, given enough of them, plus interference effects, presumably one might achieve this.

In spite of these difficulties and the additional problem that we are still in a region of resonances, we return to the Regge analysis. Zweig [22]

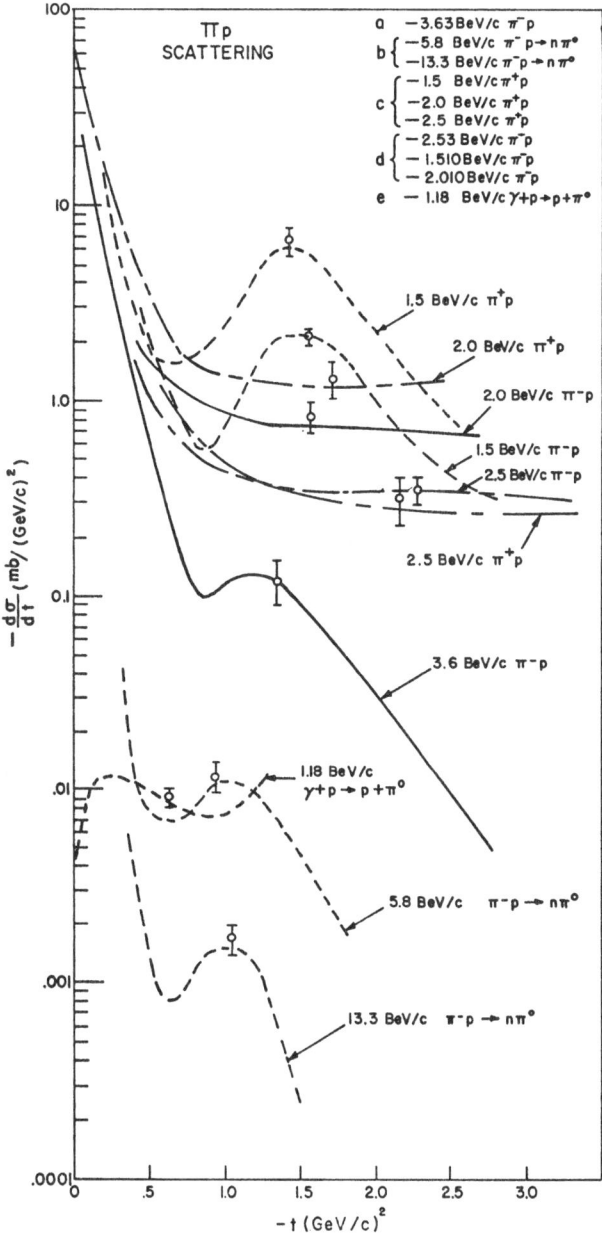

Fig. 6. Data from various π N scattering experiments [34—37] and one π-photoproduction experiment [15]

7*

has shown that the photoproduction cross-section should behave as

$$\frac{d\sigma}{d\Omega} = F(t)\, s^{2\alpha(t)-1}$$

considering only one trajectory. The most plausible one for high-momentum transfers would be ω^0 and/or ρ, since these trajectories lie higher

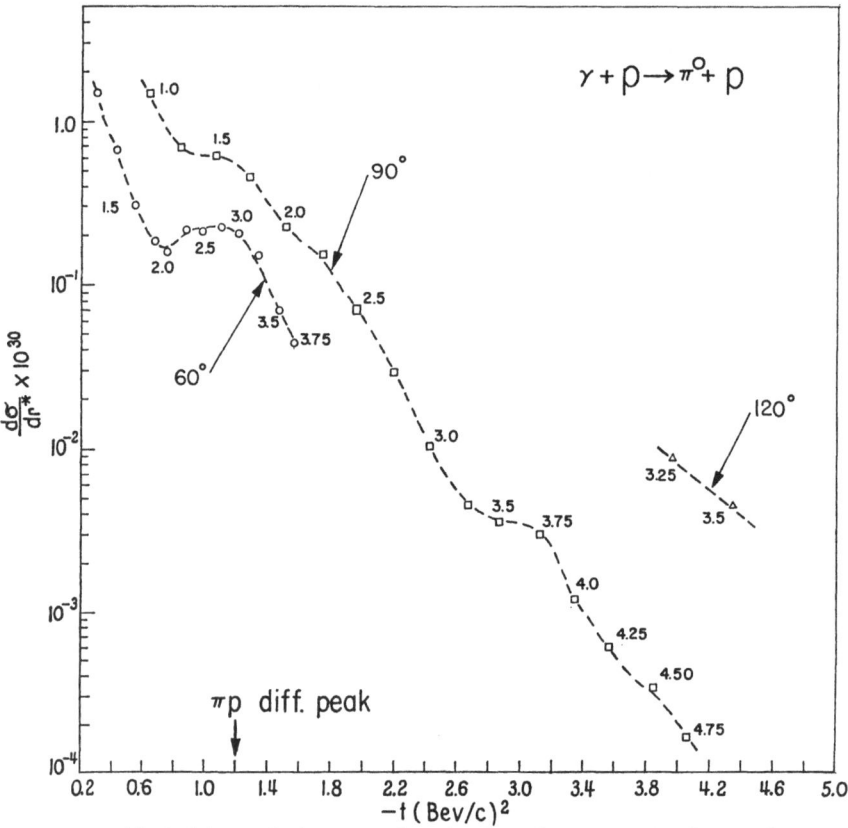

Fig. 7. π^0 photoproduction cross-sections, plotted vs t, the momentum transfer squared

than the π trajectory. By taking points from the 60° and 90° curves of Fig. 7 at a given t, we obtain

$$\frac{d\sigma/d\Omega\,(90°)}{d\sigma/d\Omega\,(60°)} = \left[\frac{s\,(90°)}{s\,(60°)}\right]^{2\alpha(t)-1}.$$

The values obtained for $\alpha(t)$ are shown in Fig. 8 together with a sketch of the presumed ρ and ω trajectories; the trajectories shown correspond to those deduced from π charge-exchange experiments [27]. Independent of Regge formalism one can say that at constant momentum transfer, the cross-section falls faster with energy the higher one chooses the momentum transfer.

At low energies the π-exchange term in photoproduction gives rise to the well known field-theoretic term with a pion propagator

$[\varDelta_\pi^2 - \mu^2]^{-1}$; this term gives rise to the pole at the unphysical angle $\cos\theta_\pi = 1/\beta_\pi$. In an analogous way there is also the term (required by gauge invariance) where the nucleon is "propagated" giving a $[\varDelta_n^2 - M^2]^{-1}$ term and this gives rise to a pole at $\cos\theta_\pi = -1/\beta_n$. At low

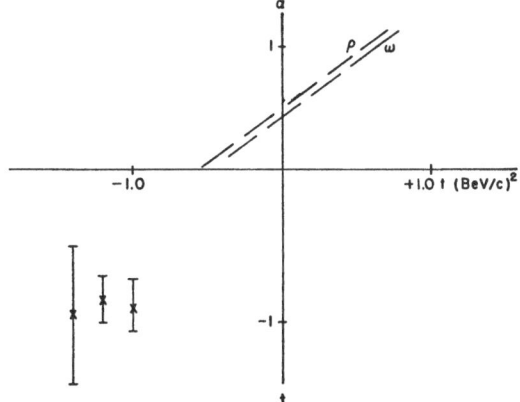

Fig. 8. Values of Regge trajectory parameter, α, derived from π^0 photoproduction

energies $\varDelta_n^2 \ll M^2$ and this term boils down to the $\sigma \cdot \varepsilon$ term responsible for S-wave π^+ photoproduction at threshold. At higher energies both terms presumably still exist though moderated by Reggeiazation, form factorization, or what have you. Nevertheless one might expect a peaking in the cross-section for backward-photoproduced π mesons or backward-scattered π mesons, and in Regge language this would correspond to the exchange of a particle corresponding to any or all of the four N* trajectories ($T = 1/2, 3/2$; odd and even).

A theoretical calculation of cross-sections for N* exchange is even less reliable than for boson exchange. One can consider that the nucleon pole is still far from the physical region or, analogously, one is extrapolating the nucleon Regge trajectories from the region of $+M^2$ to $-M^2$, a mighty jump. Nevertheless, a peak in the backward direction might be expected and is indeed found in π scattering [28] *. We have made some measurements of π^0 production at 120° (see Fig. 4), and we note that here, also, the cross-section rises above that at 90°, which indicates that the nucleon-exchange mechanism may indeed be present.

Another mechanism for backward peaks arises in resonance production since, at high J values, the angular distribution peaks near or at 0° and 180°. However, our 120° points show no spectacular energy variation.

V. $SU(3)$ and Photoproduction

One of the most interesting aspects of photoproduction experiments concerns the role of the photon in $SU(3)$. Levinson, Lipkin, and

---

* See Fig. 10 on p. 16.

Meshkov [29] have noted that if one requantizes an $SU(3)$ group such that particles of the same charge are grouped together into sub-multiplets labeled by a new quantum number, U, then one forms, for example, $\Sigma^+ p^+$ in a $U = 1/2$ group, etc. The eigenfunctions of U-spin for $U = 1$ and $U = 0$ are, however, n, $\frac{1}{2}\Sigma^0 + \frac{\sqrt{3}}{2}\Lambda^0$, Ξ^0 and $\frac{\sqrt{3}}{2}\Sigma^0 - \frac{1}{2}\Lambda^0$ respectively. Since each U-plet interacts the same way with the electromagnetic field (or so we suppose) then the photon has a U-spin of 0.

Given conservation of U-spin we predict relations between photoproduction matrix elements:

$$\left.\begin{aligned} M_a &: \gamma + p \to N^{*-} + \pi^+ + \pi^+ \\ M_b &: \gamma + p \to Y_1^* + K^+ + \pi^+ \end{aligned}\right\} \quad \left|\frac{M_a}{M_b}\right|^2 = \frac{3}{2} \tag{1}$$

$$\left.\begin{aligned} M_c &: \gamma + p \to N^{*0} + \pi^+ \\ M_d &: \gamma + p \to Y_1^{*0} + K^+ \end{aligned}\right\} \quad \left|\frac{M_c}{M_d}\right|^2 = 2 \tag{2}$$

$$\left.\begin{aligned} M_e &: \gamma + p \to n + \pi^+ \\ M_f &: \gamma + p \to \Sigma^0 + K^+ \\ M_g &: \gamma + p \to \Lambda^0 + K^+ \end{aligned}\right\} \quad M_e = -\frac{1}{\sqrt{2}}M_f - \sqrt{\frac{3}{2}}\,M_g \tag{3}$$

which gives a triangular inequality among the three cross-sections.

The first two relations have been examined by the Cambridge Bubble Chamber Group [30]. They obtain 0.8 ± 0.5 for the first and 2.0 ± 1.7 for the second which is not in disagreement with predictions.

At high energies one has the hope that kinematic corrections need not be important since in this region particle mass differences become small compared to Q-values.

Our group has measured the cross-sections e, f, and g, at 3.7 GeV and $\theta_{cm} = 45°$.

We chose this bombarding energy because of the formation of the $N^*_{3/2}$ (2750). A decuplet decays equally into $\Sigma^0 K^+$ and $n \pi^+$; this is probably a more general result since the ΣK and $n \pi$ exhibit the same positions in their octets with the role of baryon and meson interchanged; thus, from a group theoretical point of view the decays are the same to within a $\pm$ sign.

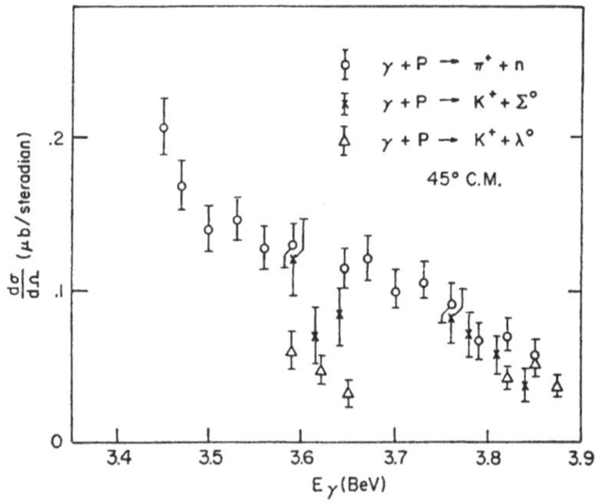

The results are shown in Fig. 9.

We measure the K^+ yields for a given electron energy. The K^+'s in the top 3% momentum bins can only be due to $K^+ \Lambda^0$ reactions;

CENTER OF MASS DIFFERENTIAL CROSS SECTION
vs
INCIDENT GAMMA ENERGY

Fig. 9. Comparison of $K^+ \Lambda^0$, $K^+ \Sigma^0$, π^+ n photoproduction cross-sections at 3.7 GeV and $\theta = 45°$

we then ran the machine energy lower and repeated. Subtracting the now known $K^+ \Lambda^0$ yields, the top 3% of the remainder can come only from $K^+ \Sigma^0$. Corrections have been made for decay in flight and nuclear absorption in the gas Cerenkov counter. We note that $\sigma(K^+ \Sigma^0) \approx$ $\approx \sigma(\pi^+ n)$ in the resonance region.

Furthermore the $K^+ \Lambda^0$ cross-section is comparable with the other two so that the triangular inequalities mentioned above are easily satisfied. Since we had expected K^+ cross-sections to be smaller than

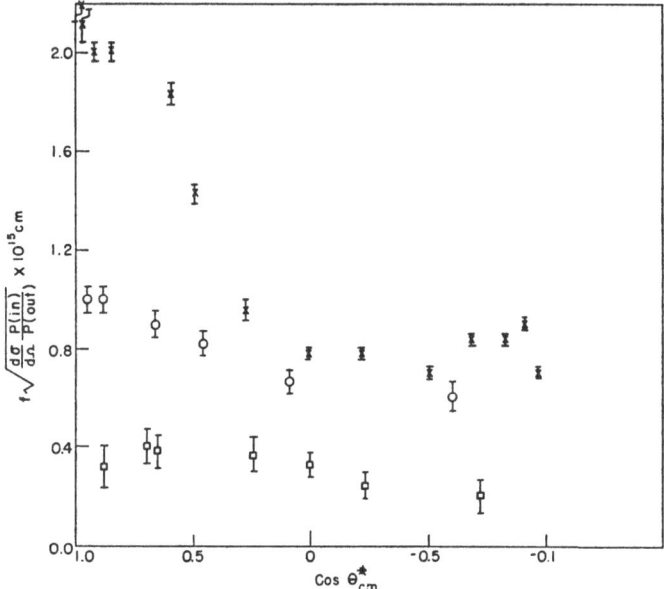

Fig. 10. Comparison of $K^+ \Lambda^0$, $K^+ \Sigma^0$, and $\pi^+ n$ photoproduction matrix elements at 1.2 GeV. $\times \pi^+ n (f = 1$, 1200 MeV, ref. [33]); $\circ$ $K^+ \Lambda (f = \sqrt{3/2}$, 1200 MeV, ref. [32]); $\square$ $K^+ \Sigma^0 (f = 1/\sqrt{2}$, 1157 MeV, ref. [31])

π^+ cross-sections, this result surprised us. We turned back to the pioneering work done at CAL. TECH. and CORNELL [31−33] to check consistency with $SU(3)$. These cross-sections (Fig. 10) have been corrected for phase space and their square roots taken to give a measure of $M_{e, f, g}$. We expect

$$\sqrt{2} \, |M_e(\pi^+ n)| \leqq |M_f(K^+ \Sigma^0)| + \sqrt{3} \, |M_g(\Lambda^0 K^+)| \,.$$

Indeed the inequality is satisfied except at small angles. However, it is clear that production is taking place in other than S-states; therefore, a proper correction to compare matrix elements should involve some higher power of the meson momentum. This would probably be sufficient to get agreement. We conclude, therefore, that $SU(3)$ predictions appear to hold in photoproduction reactions or, more modestly, that U-spin is conserved, as far as the measurements mentioned above are concerned.

These results raise more questions. Since one would not expect π^+ and K^+ angular distributions to be identical, particularly in the peripheral region, where different mass particles are being exchanged, one could expect a breakdown of relation (3) at small angles. We plan to measure this.

Finally, we can mention an interesting possibility if nature provided an $SU(3)$, $\overline{10}$. The excited proton corresponds to a U-spin $= 3/2$ state; since the proton has U-spin $= 1/2$, a photon could not excite this state. However, both neutron and excited neutron are members of U-spin $= 1$ states. Thus one should be watchful for any resonances not excited by γ p reactions though excited by γ n reactions.

References

[1] VAN HOVE, L.: p. 1 of this volume
[2] DRELL, S. D.: Rev. Modern Phys. 33, 458 (1961) and p. 71 of this volume
[3] BLUMENTHAL, R. B., W. L. FAISSLER, P. M. JOSEPH, L. J. LANZEROTTI, F. M. PIPKIN, D. G. STAIRS, J. BALLAM, H. DE STAEBLER JR., and A. ODIAN: Phys. Rev. Letters 11, 496 (1963)
[4] BLANPIED, W. A., J. S. GREENBERG, V. W. HUGHES, D. C. LU, and R. C. MINEHART: Phys. Rev. Letters 11, 477 (1963)
[5] See reports of Cambridge Bubble Chamber Group, Plenary Session I, this conference
[6] LANZEROTTI, L. J., R. B. BLUMENTHAL, D. C. EHN, W. L. FAISSLER, P. M. JOSEPH, F. M. PIPKIN, J. K. RANDOLPH, J. J. RUSSELL, J. TENENBAUM, and D. G. STAIRS: Parallel Session 1 B, this conference
[7] BLANPIED, W. A., J. S. GREENBERG, V. W. HUGHES, P. KITCHING, D. C. LU, and R. C. MINEHART: Phys. Rev. Letters 14, 741 (1965)
[8] The authors involved in various stages of the work of the M. I. T. photoproduction group are: R. ALVAREZ, Z. BAR-YAM, V. ELINGS, R. FESSEL (CEA), D. GARELICK, S. HOMMA, W. KERN, R. LEWIS, D. LUCKEY, L. OSBORNE, S. TAZZARI, and J. UGLUM. This work is supported by AEC Contract AT(30-1)-2098
[9] ALVAREZ, R., Z. BAR-YAM, W. KERN, D. LUCKEY, L. S. OSBORNE, S. TAZZARI, and R. FESSEL: Phys. Rev. Letters 12, 707 (1964)
[10] CITRON, A., W. GALBRAITH, T. F. KYCIA, B. A. LEONTIC, R. H. PHILLIPS, and A. ROUSSET: Phys. Rev. Letters 13, 205 (1964)
[11] This counter is described in detail in a paper by D. LUCKEY and R. LEWIS, Parallel Session 3 A of this conference
[12] DIDDENS, A. N., E. W. JENKINS, T. F. KYCIA, and K. F. RILEY: Phys. Rev. Letters 10, 262 (1963)
[13] CLARK, W. A., and C. E. MOLNAR: Computers in Biomedical Research, Chapter II. New York: Academic Press 1965
[14] HÖHLER, G., and J. GIESECKE: Phys. Letters 12, 149 (1964)
— 50th National Congress of the Italian Physical Society (1964); to be published in Supplement to Il Nuovo Cimento
[15] TALMAN, R. M., C. R. CLINESMITH, R. GOMEZ, and A. V. TOLLESTRUP: Phys. Rev. Letters 9, 177 (1962)
[16] For example, E. FERRARI and F. SELLERI: Nuovo cimento Suppl. 24, 453 (1962)
[17] JACKSON, J. D.: Rev. Modern Phys. 37, 484 (1965); this gives other references to this literature
[18] ROGERS, A.: (thesis) Mass. Inst. of Tech. (1965), and Session I, this conference (theoretical calculations by J. D. JACKSON on the absorption model are included in this thesis)
[19] BERMAN, S. M., and S. D. DRELL: Phys. Rev. 133, B 791 (1964)
[20] KRAMER, G., and P. STICHEL: Z. Physik 178, 519 (1964)
[21] CHILDERS, R. W., and W. G. HOLLADAY: Phys. Rev. 132, 1809 (1963)

[22] ZWEIG, G.: Calif. Inst. of Tech., Synchrotron Lab., Report, CTSL-40. Pasadena, August 1963
[23] BRONZAN, J. B., and F. E. LOW: Phys. Rev. Letters 12, 522 (1964)
[24] DAMOUTH, D. E., L. W. JONES, and M. L. PERL: Phys. Rev. Letters 11, 287 (1963)
[25] SIMMONS, L. M.: Phys. Rev. Letters 12, 229 (1964)
[26] For example, D. BARGE, W. CHU, L. LEIPUNER, R. CRITTENDEN, H. J. MARTIN, F. AYER, L. MARSHALL, A. C. LI, W. KERNAN, and M. L. STEVENSON: Phys. Rev. Letters 13, 69 (1964); or A. DAR, M. KUGLER, Y. DATHAN, and S. NUSSINOV: Phys. Rev. Letters 12, 82 (1964)
[27] LOGAN, R.: Phys. Rev. Letters 14, 414 (1965)
[28] Aachen-Berlin-Birmingham-Bonn-Hamburg-London(I. C.)-München Collaboration, Phys. Letters 10, 248 (1964), and R. RUBINSTEIN, J. OREAR, D. B. SCARL, D. H. WHITE, A. D. KRISCH, W. R. FRISKEN, A. C. READ, and H. RUDERMAN: Bull. Am. Phys. Soc. 10, 528 (1965)
[29] LEVINSON, C. A., H. J. LIPKIN, and S. MESHKOV: Proceedings of the International Conference on Nucleon Structure at Stanford University, 1963, p. 309. Palo Alto, California: Stanford University Press 1964
[30] CROUCH, H. R., R. HARGRAVES, B. KENDALL, R. E. LANON, A. M. SHAPIRO, M. WIDGOFF, G. E. FISCHER, A. E. BRENNER, M. E. LAW, E. E. RONAT, K. STRAUCH, J. C. STREET, J. J. SZYMANSKI, J. D. TEAL, P. BASTIEN, Y. EISENBERG, B. T. FELD, V. K. FISCHER, I. A. PLESS, A. ROYERS, C. ROYERS, C. ROSENSON, T. L. WATTS, R. K. YAMAMOTO, L. GUERRIERO, and G. A. SALANDIN: Phys. Rev. Letters 13, 636 (1964)
[31] ANDERSON, R. L., E. GABATHULER, D. JONES, B. D. McDANIEL, and A. J. SADOFF: Phys. Rev. Letters 9, 131 (1962)
[32] PECK, C. W.: Phys. Rev. 135, B 830 (1964)
[33] KILNER, J. R.: Ph. D. Thesis, Calif. Inst. of Technology 1962
[34] PERL, M. L.: S. L. A. C., Publication 58 (1964)
[35] STIRLING, A. V., P. SONDEREGGER, J. KIRZ, P. FALK-VAIRANT, O. GUISAN, C. BRUNETON, P. BORGEAUD, M. YVERT, J. P. GUILLAND, C. CAVERZASIO, and B. AMBLARD: Phys. Rev. Letters 14, 763 (1965)
[36] COOK, V., B. CORK, and W. R. HOLLEY: Phys. Rev. 130, 762 (1963)
[37] LAI, K. W., L. W. JONES, and M. L. PERL: Phys. Rev. Letters 7, 125 (1961)

Prof. Dr. L. S. OSBORNE
Synchrotron Laboratory
Massachusetts Institute of Technology
77 Massachusetts Avenue
Cambridge, Mass.

Theoretical Aspects of Colliding Beam Experiments

R. Gatto

Istituto di Fisica
dell'Università, Firenze
Laboratori Nazionali del
C. N. E. N. di Frascati, Roma

In this talk I shall try to summarize the theoretical aspects of colliding beam experiments. I shall try to be rather elementary, as this talk should be intended mostly for those who will do experiments with colliding beams.

The following table shows a list of colliding beam projects at different laboratories which are either completed or in construction or still under study.

Table 1

$e^+ e^-$				
Frascati	AdA	$e^+ e^-$	250 MeV	working
Moscow	FIAN	$e^+ e^-$	270 MeV	
Orsay	ACO	$e^+ e^-$	450 MeV	to be completed
Novosibirsk		$e^+ e^-$	700 MeV	to be completed
Frascati	Adone	$e^+ e^-$	1500 MeV	in construction
Stanford		$e^+ e^-$	3000 MeV	project
Cambridge		$e^+ e^-$	3000 MeV	study
Erevan		$e^+ e^-$	3000 MeV	study
Hamburg		$e^+ e^-$		study
$e^- e^-$				
Kharkov		$e^- e^-$	100 MeV	to be completed
Novosibirsk		$e^- e^-$	130 MeV	working
Stanford		$e^- e^-$	500 MeV	working

The topics I shall discuss are:

Electromagnetic reactions: $e^+ + e^- \to 2\gamma$, $e^+ + e^- \to e^+ + e^-$, $e^+ + e^- \to e^+ + e^- + \gamma$, $e^+ + e^- \to \mu^+ + \mu^-$, $e^- + e^- \to e^- + e^-$, $e^- + e^- \to e^- + e^- + \gamma$, etc.; vacuum polarization effects; $e^- - e^+$ annihilation into strong interacting particles; the one-photon channel; resonant contributions; annihilation into pions; intermediate ρ^0, ω^0, and φ^0; two-photon channel; annihilation into fermion-pairs; weak-vector-meson production and weak interactions.

1. Colliding $e^- - e^-$ and $e^+ - e^-$ beams can be used to test electrodynamics. Unfortunately for an evaluation of the tests one would need a model for the breakdown. The usual procedure of modifying the propagators [1] has a great disadvantage in its lack of gauge-invariance.

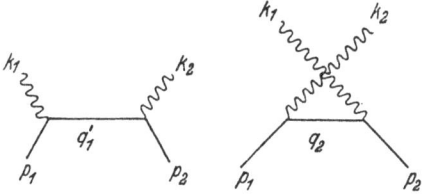

Fig. 1. Graphs for $e^+ + e^- \rightarrow 2\gamma$

A recent paper by DRELL and McCLURE [2] discusses a better model of breaking but contains more form factors. In view of these facts our considerations on tests of quantum electrodynamics will be openly superficial and only of indicative value.

I shall first discuss the electromagnetic reactions

$$e^+ + e^- \rightarrow 2\gamma \tag{1}$$

$$e^+ + e^- \rightarrow e^+ + e^- \tag{2}$$

$$e^+ + e^- \rightarrow e^+ + e^- + \gamma \tag{3}$$

$$e^+ + e^- \rightarrow \mu^+ + \mu^- . \tag{4}$$

2. Reaction (1) occurs in lowest order through the graphs (Fig. 1) In c. m.

$$q_1^2 = 4E^2 \sin^2 \frac{\theta}{2}$$

$$q_2^2 = 4E^2 \cos^2 \frac{\theta}{2}$$

where E is the c. m. energy of e^- (or e^+) and θ is the annihilation angle. In the relativistic limit and for $\theta \gg (m_e/E)$

$$\frac{d\sigma}{d(\cos\theta)} = (\pi \, r_0^2) \left(\frac{m_e}{E}\right)^2 \frac{|F(q_1^2)|^2 \cos^4 \frac{\theta}{2} + |F(q_2^2)|^2 \sin^4 \frac{\theta}{2}}{\sin^2\theta} \tag{5}$$

where r_0 is the electron radius and we have introduced a form factor $F(q^2)$, depending on the transferred momentum, to account for possible breaking of the theory. The form factor F can be thought as a product, $F = V_e^2 \, P_e$, of a form factor for the electron vertex and a form factor for the electron propagator. If $F = 1$ from (5) one gets

$$\frac{d\sigma}{d(\cos\theta)} = (\pi \, r_0^2) \left(\frac{m_e}{E}\right)^2 \frac{1}{2} \frac{2 - \sin^2\theta}{\sin^2\theta} \tag{6}$$

which holds, like (5) for $\theta \gg (m_e/E)$. For θ near $0°$

$$\frac{d\sigma}{d(\cos\theta)} = (\pi \, r_0^2) \left(\frac{m_e}{E}\right)^2 \frac{1}{\sin^2\theta + \left(\dfrac{m_e}{E}\right)^2 \cos^2\theta} \, . \tag{7}$$

We have plotted in Fig. 2 the differential cross-section for various energies E. The total cross-section is given by

$$\sigma = (\pi \, r_0^2) \left(\frac{m_e}{E}\right)^2 \left(\log \frac{2E}{m_e} - \frac{1}{2}\right) . \tag{8}$$

Let us take

$$F(q^2) = \frac{1}{1 + \dfrac{q^2}{Q^2}} \, . \tag{9}$$

One finds for $Q = 1 \text{ GeV}$ (corresponding to 0.2 fermi) that the cross-section at $E = 250 \text{ MeV}$ is changed from its value with $F = 1$ as follows:

$$\begin{array}{ll} \text{at } 90° & \text{by} - 12\% \\ 60° & \quad - 7\% \\ 45° & \quad - 4\% \\ 30° & \quad -0.7\% \, . \end{array}$$

Thus, very roughly, at $E = 250 \text{ MeV}$, if one can distinguish a 7% effect at $60°$ one can test the theory to 0.2 fermi. If $E = 500 \text{ MeV}$ a similar

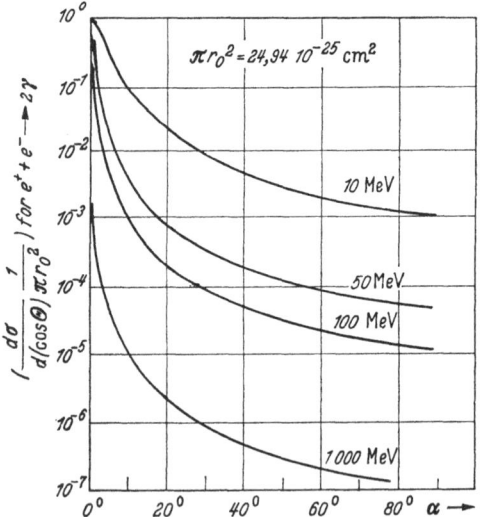

Fig. 2. Differential cross-section of $e^+ + e^- \rightarrow 2\gamma$ in lowest order

experiment would test the theory to 0.1 fermi, according to this model where the effect depends on $(E/Q)^2$. Similarly if $E = 1 \text{ BeV}$ the test is up to 0.05 fermi. The use of reaction (1) as a monitoring process for beam collisions is unfortunately not convenient. The monitoring process must be such that in the particular geometry it involves only very small momentum transfers even at the highest energy. It will then be very reliably predicted by the theory, and from the calculated cross-section one can check the beam dimensions by a uniform and instantaneous procedure. The 2γ annihilation at forward angles would be a possible monitor except that its cross-section is too low in that region in comparison to other processes that show a stronger singularity in the forward direction. In particular the process

$$e^+ + e^- \to e^+ + e^- + 2\gamma \tag{10}$$

with the photons going in opposite directions is expected to contaminate the measurement of 2γ annihilation. BAYER and GALITSKY [3] have calculated the cross-section of (10) in a classical approximation. The cross-section of double bremsstrahlung is written as

$$d\sigma_2 = d\sigma_0 \frac{1}{2!} \, dw(k_1) \, dw(k_2) \tag{11}$$

where

$$dw = \frac{1}{2\omega} j_\mu j^\mu \, d^3k \tag{12}$$

and

$$j_\mu = \frac{i\,e}{(2\pi)^{3/2}} \left[\sum_i \frac{(p_i)_\mu}{(k\,p_i)} - \sum_j \frac{(p'_j)_\mu}{(k\,p'_j)} \right]. \tag{13}$$

p_i being the momentum of an incoming particle and p'_j that of an outgoing particle [4]. For the case considered, BAYER and GALITSKY give the result (in c. m.)

$$\frac{d\sigma(e^+ + e^- \to e^+ + e^- + 2\gamma)}{d\sigma(e^+ + e^- \to 2\gamma)} = \frac{2.3 \times 32 \, \alpha^2 \, \gamma^2}{\pi^2 \ln 2\gamma} \frac{d\omega_1}{\omega_1} \frac{d\omega_2}{\omega_2} \tag{14}$$

where $\gamma = E/m_e$, and ω_1 and ω_2 are the photon energies. Eq. (14) shows that 2γ bremsstrahlung will be larger than 2γ annihilation at sufficiently high energy, in the approximation used. A complete calculation of 2γ bremsstrahlung, including a complete calculation of hard photon emission, has not however been made so far. The radiative corrections of $e^+ + e^- \to 2\gamma$ have been recently calculated by TSAI [5]. TSAI is interested in the possible use of $e^+ + e^- \to 2\gamma$ to produce high energy γ-rays. He therefore calculates the spreading of the photon peak at a fixed angle in positron-hydrogen-atom collisions, which arises from radiative corrections. The radiative corrections obtained by TSAI do not contain dangerous terms of the kind $(\alpha/\pi) \ln^2(4E^2/m_e^2)$ which cancel out after addition of the elastic and the inelastic contributions. Such $\ln^2$-terms had been found in previous calculations but their appearance was due to the unnatural choice of phase-space for the additional bremsstrahlung photon which had been supposed to include energies ω_3 up to a maximum value $\ll m_e$ and isotropic in the laboratory system. Terms of the kind $(\alpha/\pi) \ln^2(4E^2/m_e^2)$ would become of the order of unity at energies of the order of a few BeV thus making the whole perturbative approach invalid, for instance when interpreting a high energy $e^+ - e^-$ colliding beam experiment. The phase-space for the bremsstrahlung photon is dictated by the experimental set-up of the particular experiment. TSAI gives for the radiative correction δ, defined by

$$\frac{d\sigma}{d\Omega} = \frac{d\sigma_0}{d\Omega}(1+\delta) \tag{15}$$

the expression

$$\delta = -\frac{2\alpha}{\pi}\left\{\left(\ln\frac{E}{K_3^{\max}} - \frac{3}{4}\right)\left(\ln\frac{4E^2}{m_e^2} - 1\right) + \frac{f}{2}\right\} \tag{16}$$

where f is a complicated function, usually rather small in the experimental conditions. The cut-off $K_3^{\max}$ is to be calculated from the energy and angular resolution of the detectors. Eq. (16) has a striking similarity to Schwinger's radiative corrections to potential scattering.

$$\delta_{\text{Schwinger}} \approx -\frac{2\alpha}{\pi}\left\{\left(\ln\frac{\mathscr{E}}{\Delta\mathscr{E}} - \frac{13}{12}\right)\left(\ln\left(\frac{-q^2}{m_e^2}\right) - 1\right) + \frac{17}{36}\right\} \tag{17}$$

provided one substitutes

$$\frac{\mathscr{E}}{\Delta\mathscr{E}} \to \frac{E}{K_3^{\max}} \quad \text{and} \quad -q^2 \to 4E^2.$$

TSAI notices that the two expressions become almost identical if one subtracts the vacuum polarization contribution from Schwinger's δ, corresponding to the fact that there are no such contributions in the annihilation problem. A calculation of the radiative corrections to the total $e^+ + e^- \to 2\gamma$ cross-section has also been performed by ANDREASSI et al. [6]. Actually ANDREASSI et al. calculate the total cross-section for $e^+ - e^-$ annihilation into photons to order e^6 (i. e. annihilation into 2γ and 3γ)

$$\sigma_{\text{tot}} = \sigma_D(1+\delta_T)$$

where σ_D is the Dirac $e^+ + e^- \to 2\gamma$ cross-section. In the E. R. limit they find

$$\delta_T = \frac{\alpha}{12\pi}\left[2\ln^2 2\gamma - \ln 2\gamma + (4\pi^2 - 13) - \frac{5\pi^2 - 11}{\ln 2\gamma - 1}\right].$$

The presence of the $\ln^2(2\gamma)$ term is the most significant result. A theorem by ERIKSSON and PETERMAN [7] says that the radiative corrections to a differential cross-section for large values of the squared momentum transfer in c. m. (calculated from soft photons) are expandible in a power series of $(\alpha/\pi)\ln(q^2/m^2)$. Specifically for $q^2 \gg m^2$ in c. m., to order e^6, according to this theorem one expects

$$\delta = \frac{\alpha}{\pi}\left[c_1\ln\frac{E}{\Delta E}\ln\frac{q^2}{m^2} + c_2\ln\frac{q^2}{m^2} + c_3\right]$$

where c_1, c_2, c_3 are angle-dependent. Addition of hard photons may add terms $\ln^2(E/\Delta E)$ but not modify the main conclusion. According to ANDREASSI, BUDINI and FURLAN the $\ln^2(2\gamma)$ comes from the region of small q^2 (where the theorem does not hold) and precisely from soft photons.

3. The reaction (2), electron-positron scattering, is obtained in lowest order from the graphs of Fig. 3. In c. m.

$$q_1^2 = 4 E^2 \sin^2 \frac{\theta}{2}$$

$$K^2 = -4 E^2$$

(q_1 is spacelike, K is timelike). For angles $\theta \neq 0$ and in the relativistic limit

$$\frac{d\sigma}{d(\cos\theta)} = (\pi r_0)^2 \left(\frac{m_e}{E}\right)^2 \left\{ \frac{1}{4} \frac{1 + \cos^4 \frac{\theta}{2}}{\sin^4 \frac{\theta}{2}} |F(q_1^2)|^2 - \frac{1}{2} \frac{\cos^4 \frac{\theta}{2}}{\sin^2 \frac{\theta}{2}} \times \right.$$

$$\left. \times \mathrm{Re}\,[F(q_1^2) F^*(K^2)] + \frac{1}{8} (1 + \cos^2\theta) |F(K^2)|^2 \right\} \tag{18}$$

We have introduced a form factor F which can be thought of as a product $V_e^2 P_\gamma$ of the electron vertex squared and of the photon propagator.

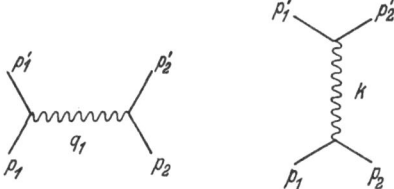

Fig. 3. Graphs for $e^+ + e^- \to e^+ + e^-$

This time V_e refers to a situation where the electron lines are on the mass-shell and the photon is off-mass-shell. On the other hand in the 2γ annihilation of Fig. 1 the incident electron and the photon were on mass-shell and the final electron was off-mass-shell. The two form-factors could then strictly be different. Graphs for $F = 1$ are reported in Fig. 4. At small angle the cross-section arises from transfer of very small momentum. In fact the annihilation graph in Fig. 3 is not singular in the forward direction. The scattering graph dominates in that region and the transferred momentum $q_1^2 \to 0$. Unfortunately, the use of $e^+ - e^-$ scattering as a monitoring process is made difficult just by the singularity of the cross-section at $\theta = 0°$. The dependence on the geometry of the apparatus would be too critical because of the rapid variation of the cross-section. At larger angles, $e^+ - e^-$ scattering can be measured and used for testing the theory. We have reported in Fig. 5 the relative correction to the differential $e^+ - e^-$ cross-section calculated from Eq.(18) with

$$F(q^2) = \frac{1}{1 + \dfrac{q^2}{Q^2}}$$

for $Q^{-1} = 0.1$ fermi and $Q^{-1} = 0.3$ fermi and for various energy. With a 10% experiment one can measure an effect for $Q^{-1} = 0.3$ fermi

$(Q = 670 \text{ MeV})$ provided one goes at an energy $E \gtrsim 150 \text{ MeV}$. With the same 10% precision if $Q^{-1} = 0.1$ fermi $(Q = 2 \text{ GeV})$ one has to go to $E \gtrsim 350 \text{ MeV}$. With the above choice of $F(q^2)$ the net effect results in a decrease of the cross-section. The maximum deviation would be observed around 90%.

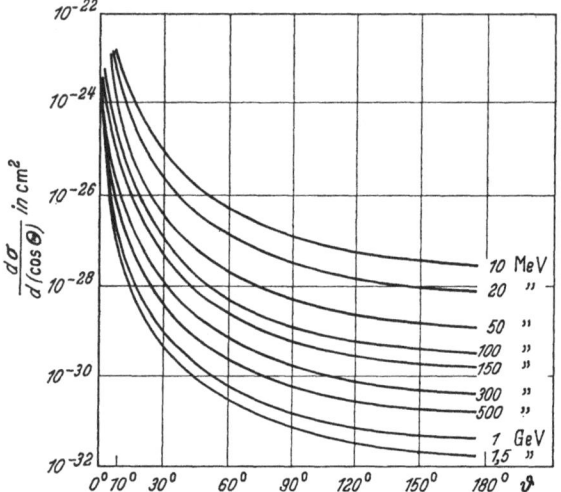

Fig. 4. Differential cross-section for $e^+ + e^- \rightarrow e^+ + e^-$ at lowest order

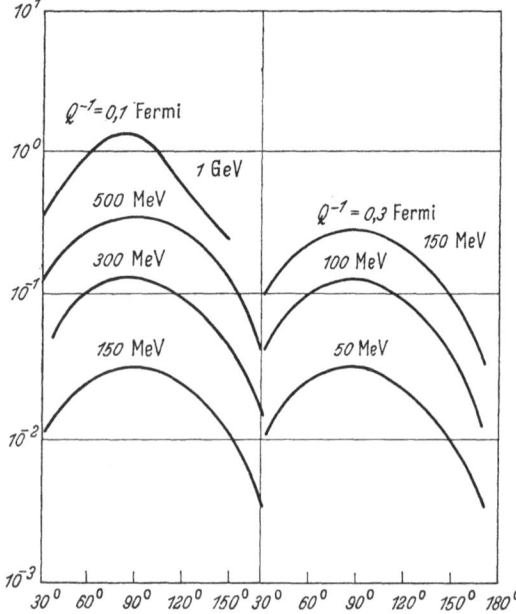

Fig. 5. Relative correction to the differential cross-section for $e^+ + e^- \rightarrow e^+ + e^-$ from a simple model of breakdown

4. The reaction (3), $e^+ + e^- \rightarrow e^+ + e^- + \gamma$, is the best candidate as a monitoring reaction. As difficulty with such a choice has been considered the fact that the beam can produce γ-rays also by interaction with the residual gas and it would be difficult to separate the two sources on the basis of the shapes of the γ-spectra. However the AdA group [8] has examined the problem and concluded that the separation is indeed possible and poses no relevant problem with machines larger than AdA.

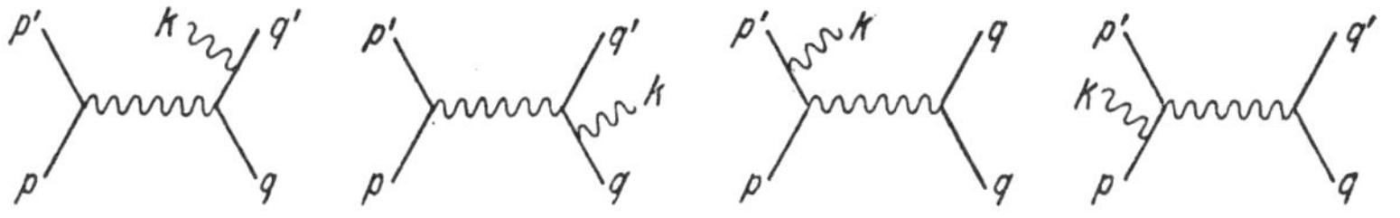

Fig. 6a. Graphs for $e^+ + e^- \rightarrow e^+ + e^- + \gamma$: scattering graphs

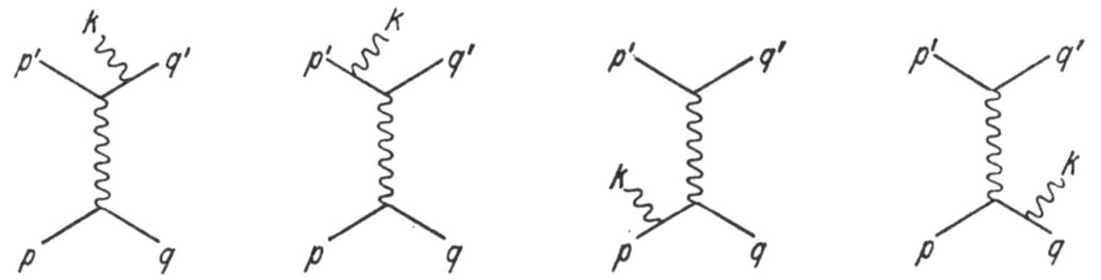

Fig. 6b. Graphs for $e^+ + e^- \rightarrow e^+ + e^- + \gamma$: annihilation graphs

The theoretical study of single bremsstrahlung in $e^+ - e^-$ collisions is also of relevance for computing the radiative corrections to $e^+ + e^- \rightarrow \rightarrow e^+ + e^-$. An evaluation of the radiative corrections of $e^+ + e^- \rightarrow \rightarrow e^+ + e^-$ implies the integration of the hard photon emission cross-section for those photons which escape detection in the particular experimental situation. There are eight Feynman diagrams for single bremsstrahlung which are reported in Fig. 6a and in Fig. 6b. We call the graphs of Fig. 6a scattering graphs and those of Fig. 6b the annihilation graphs. Single bremsstrahlung calculations have been performed by GARIBYAN [9] and more recently by ALTARELLI and BUCCELLA [10]. These last authors calculate the angular distributions of the photons around the direction of the incident particles after integrating on the final electrons. They also calculate the energy distribution of the emitted radiation and the integrated spectrum from a given c. m. cut-off energy.

Particularly powerful approximations can be made in the region of high energies and small angles of the emitted photon with respect to the colliding beam direction. The annihilation graphs can be neglected in comparison with the scattering graphs on the basis that the denominator of the photon propagator is much smaller for a scattering graph than for an annihilation graph. In a scattering graph the denominator is $(p - p')^2$ or $(q - q')^2$, which for small angle becomes $\sim 4 m_e^6 K_0^2/W^6$ where W is the total c. m. energy and K_0 is the emitted photon energy. On the other hand in an annihilation graph the denominator is $(p + q)^2 = W^2$ or $(p' + q')^2 = W^2 - 2 W K_0$ and, excluding a small region near the maximum photon energy $(K_{0\,max} \sim W/2)$, where however the cross-section approaches zero, these denominators are much larger by many orders of magnitudes than those of the scattering graphs. By a different reason one can neglect the interferences between the first two and the last two graphs of Fig. 6a. In fact the radiation emitted from the electron is contained in a half-cone around the electron line, with an opening angle $\sim m_e/p_0$. The radiation emitted from the positron similarly is in a narrow cone around the positron. There is practically no overlap between the two cones in the most frequent situation of scattering at small angle.

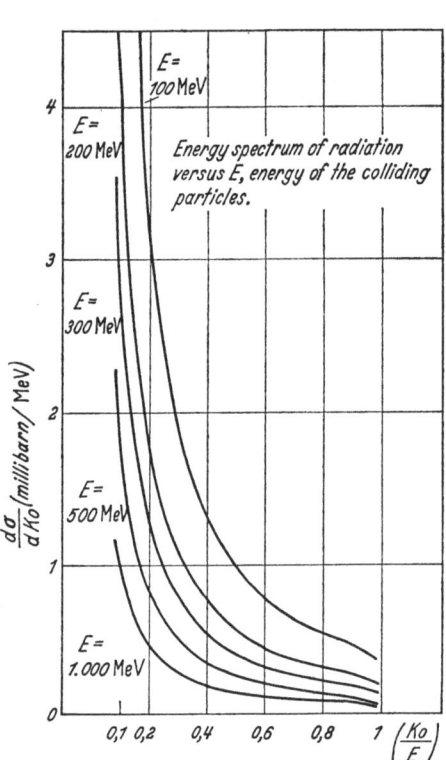

Fig. 7. Bremsstrahlung spectrum for $e^+ + e^- \rightarrow e^+ + e^- + \gamma$

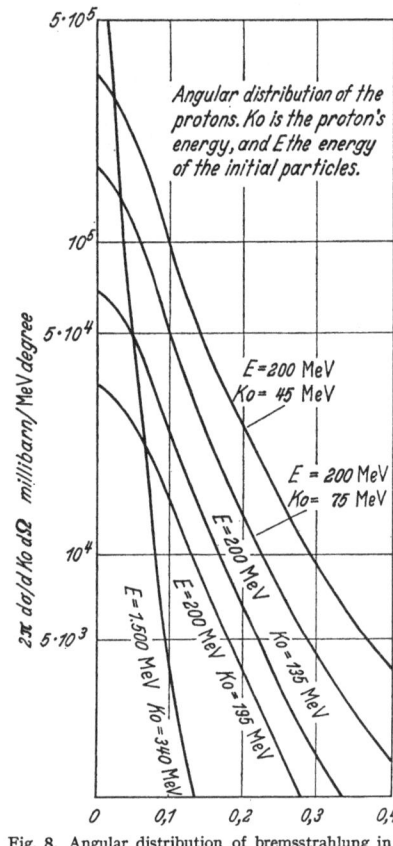

Fig. 8. Angular distribution of bremsstrahlung in $e^+ + e^- \rightarrow e^+ + e^- + \gamma$

the most frequent situation of scattering at small angle. Of course, hard photon emission with final electrons at large angles requires a different calculation, as the approximations made here, of neglecting the annihilation graphs, and of neglecting the interference between positron emission and electron emission would have to be improved. In Fig. 8 we report the angular distributions of the photons. The quantity $(2\pi \, d\sigma)/(dK_0 \, d\Omega)$ is reported in (millibarn)/(BeV) × (degree) as a function of the angle between the final photon and the initial positron for $E = 200$ MeV (and various values of K_0) and for $E = 1500$ MeV and $K_0 = 240$ MeV. The energy spectrum $d\sigma/dK_0$ can be expressed by

$$\frac{d\sigma}{dK_0} \cong 4\alpha \, r_0^2 \frac{1}{K_0} \left[\frac{2(W^4 + \varrho^4)}{W^4} - \frac{4}{3} \frac{\varrho^2}{W^2} \right] \left[\log \frac{\varrho^2}{2m_e K_0} \frac{W}{m_e} - \frac{1}{2} \right] \quad (19)$$

where r_0 is the electron radius, $W^2 = (p + q)^2$, $\varrho^2 = (p' + q')^2 = W^2 - 2W K_0$ and K_0 is the photon energy. The singularity of the photon spectrum at small K_0 is higher than $1/K_0$ reflecting the singularity of the original process. The integrated spectrum is given by

$$\sigma(\varepsilon) \cong 4\,\alpha\,r_0^2 \left\{ \left(\frac{8}{3} \log \frac{W}{2\varepsilon} - \frac{5}{3} \right) \times \right.$$
$$\left. \times \left(\log \frac{W^2}{m_e^2} - \frac{1}{2} \right) + \frac{4}{3} \left[\log \frac{W}{2\varepsilon} \right]^2 - 1 - \frac{4\pi^2}{q} \right\} . \tag{20}$$

The energy spectrum for $E = 100, 200, 300, 500$ and 1000 MeV is reported in Fig. 7. The spectrum is indeed the typical bremsstrahlung spectrum in the high energy limit in a nuclear field with $Z = 1$ except for an overall factor of two, reflecting the fact that in the $e^+ - e^-$ case two particles can radiate, while in the nuclear case only the electron radiates.

5. Reaction (4), $e^+ + e^- \rightarrow \mu^+ + \mu^-$, can be useful also to test muon-structure. At lowest order the reaction proceeds through the graph of Fig. 9. The transferred four-momentum is timelike, $K^2 = -4E^2$. If the electron mass is neglected

$$\frac{d\sigma}{d(\cos\theta)} = \frac{\pi}{4}\,\alpha^2\,\lambdabar^2\,\beta_\mu \left[\frac{1}{2} (1 + \cos^2\theta) + \frac{1}{2} \left(\frac{m_\mu}{E} \right)^2 \sin^2\theta \right] F(K^2) \tag{21}$$

where $\alpha = 1/137$, $\lambdabar$ is the reduced wavelength of the incoming e^+ (or e^-), β_μ and m_μ are the velocity and mass of the final μ, and we have introduced a form factor F which can be thought of as a product of a muon vertex form-factor V_μ, an electron vertex form-factor V_e, and the photon propagator form-factor P_γ. Note that the angular dependence is not altered by the presence of the form-factor $F(K^2)$. The total cross-section is

$$\sigma = 2.18 \times 10^{-32} \text{ cm}^2 \left[\frac{1}{x^2} \left(1 - \frac{1}{x^2} \right)^{\frac{1}{2}} \left(1 + \frac{1}{2x^2} \right) |F(-4E^2)|^2 \right] \tag{22}$$

with $x = E/m_\mu$. The values of σ for $F = 1$ are reported in Fig. 10 versus x. Fig. 10 also contains the "perturbation theory" cross-sections for $e^+ + e^- \rightarrow \pi^+ + \pi^-$, $e^+ + e^- \rightarrow p + \bar{p}$, and $e^+ + e^- \rightarrow K + \bar{K}$. By "perturbation theory" cross-sections we mean the cross-section one would have for these processes if the strong-interacting particles produced would have no structure (only point-charges). These "perturbation theory" cross-sections are thus quite certainly very far from physical reality, although it may be convenient to know them for purposes of comparison. In Fig. 10 the cross-sections are reported as functions of $x = E/m$, where m is the mass

Fig. 9. Graph for $e^+ + e^- \rightarrow \mu^+ + \mu^-$

of the particle produced ($m = m_\mu, m_\pi, m_p, m_K$ respectively). To determine a breakdown of the theory in $e^+ + e^- \rightarrow \mu^+ + \mu^-$, according to the model of Eq. (21), one has to measure the absolute cross-section. The angular dependence, as we have seen, is independent of a possible breakdown

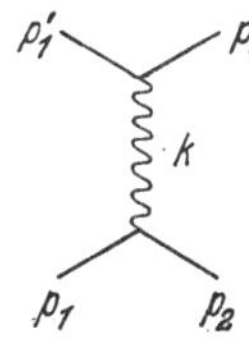

8*

as long as the one-photon channel dominates. For instance with

$$F(q^2) = \frac{1}{1 + \dfrac{q^2}{Q^2}}$$

the cross-section takes on a factor

$$|F(-4E^2)|^2 \cong 1 + \frac{8E^2}{Q^2}$$

and for $8E^2/Q^2 \sim 10\%$ one can explore up to $Q \sim 9E$, and remembering that 1 GeV corresponds roughly to 0.2 fermi one sees that for $E \sim 300$ MeV

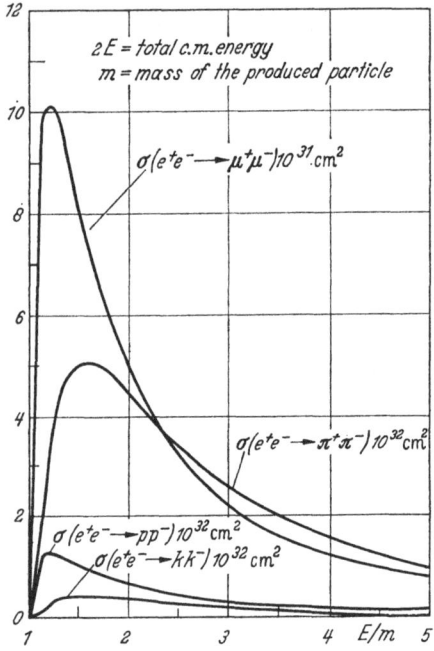

Fig. 10. "Perturbation theory" cross-sections

one can explore already up to $Q \gtrsim 0.1$ fermi and for $E \sim 1$ GeV one can explore up to $Q \gtrsim 0.03$ fermi. In view of its simpler feasibility, annihilation into a muon pair, $e^+ e^- \to \mu^+ \mu^-$, looks as one of the most promising experiments with electron-positron colliding beams. The interpretation of the experiments will require an accurate evaluation of the radiative corrections. These have been discussed by Furlan, Gatto and Longhi [11]. Hard photon emission $e^+ + e^- \to \mu^+ + \mu^- + \gamma$ has been discussed by Mosco [12] and by Longhi [13]. The graphs contributing to lowest order radiative corrections to $e^+ + e^- \to \mu^+ + \mu^-$ are reported in Fig. 11. Initial and final vertex graphs are denoted by (V_i) and (V_f) respectively. Self-energy graphs are denoted by (SE_i) and (SE_f). Bremsstrahlung graphs are denoted by (B_i) and (B_f). The two-photon-exchange graphs are called (2γ). One has to add coherently the V, SE, and 2γ graphs to the lowest order graph and then add incoherently the bremsstrahlung contributions. A great simplification occurs if one is interested in experiments that treat symmetrically the final charges (total cross-section, or any arrangement which does not distinguish between μ^+ and μ^-, etc.). In this case (experimentally the

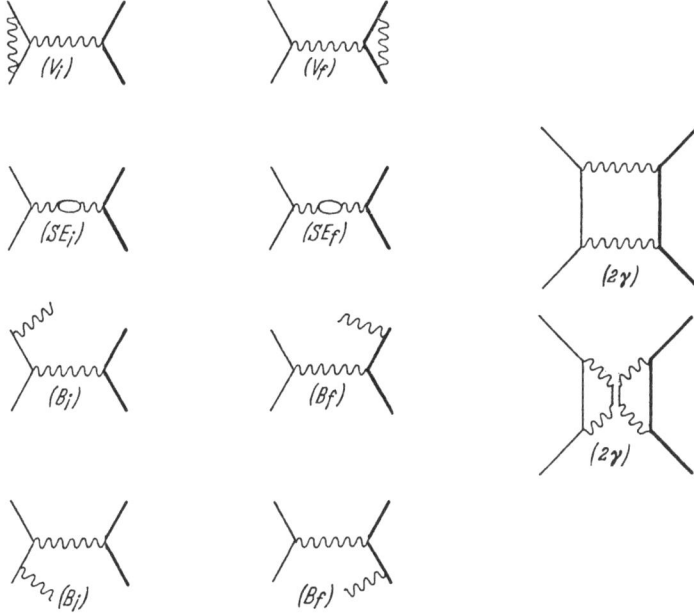

Fig. 11. Lowest order radiative corrections for $e^+ + e^- \rightarrow \mu^+ + \mu^-$

most interesting one) there is no interference between the lowest order graph (with only one photon exchanged) and the 2γ graph. Thus to the order e^6 in cross-section one can entirely neglect for such situations the 2γ graphs. Furthermore one can also neglect the interference between initial and final bremsstrahlung .The reason in both cases is the invariance under charge conjugation which does not allow a photon to transform virtually into two photons. One thus sees that one can talk consistently in these cases of an initial (interference of the lowest order graph with $V_i + SE_i$ plus the squared B_i contributions) and a final (same with i → f) radiative correction.

We shall examine here the radiative corrections assuming the following idealized experimental conditions. The muons emitted in $e^+ + e^- \rightarrow \mu^+ + \mu^-$ are observed in (approximately) opposite directions at a scattering angle θ. The detection system has an opening angle $\Delta\theta$ and the energy of a muon is observed with an energy resolution $\Delta\varepsilon$. The situation is illustrated in Fig. 12. We first look for the restrictions imposed by the energy resolution $\Delta\varepsilon$ and then for the restrictions imposed by the angular resolution of the detectors. The energy resolution $\Delta\varepsilon$

limits the energy of a bremsstrahlung photon emitted in the direction θ up to a maximum energy $\Delta\varepsilon$. For a photon emitted in the direction orthogonal to the direction θ (i. e. at angles $\theta = \pm\pi/2$) the maximum energy is instead $2\Delta\varepsilon$. In Fig. 12 we have drawed a kind of ellipsoid (with the shorter axis of $\Delta\varepsilon$ and the longer axis of $2\Delta\varepsilon$) which gives for any direction of bremsstrahlung emission the maximum energy that the

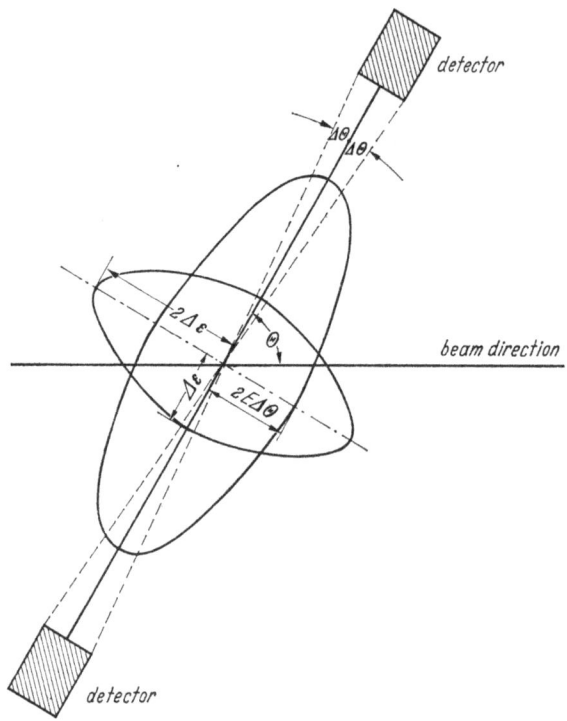

Fig. 12. Energy upper limits for a bremsstrahlung photon for an idealized detection arrangement. In an actual experiment the limits will depend on the particular detectors used

photon can take compatibly with the condition that the final muons have energy between $E - \Delta\varepsilon$ and E. Similarly one can show that the geometrical condition imposed by the detectors implies that the maximum photon energy in the direction θ is $\sim E$ and in the orthogonal direction $\sim 2E\Delta\theta$. The condition gives rise to a limitation, with varying angle, as illustrated in Fig. 12, with the kind of ellipsoid of longer axis E, and shorter axis $\sim 2E\Delta\theta$. We shall assume to be definite $\Delta\varepsilon \sim 2\%$ of E and $\Delta\theta \sim 1°$. The limit $2E\Delta\theta$, is then $\sim 3.5\%$ of E. In Fig. 12 the shadowed region indicates the possible photon momenta (the figure is rotationally invariant around the axis connecting the detectors). Under these conditions the radiative corrections are expected to be large. The physical reason is the following. The correction is almost entirely due to soft photons. The hard photons emitted from the final muons are largely cut-off from the energy resolution. Furthermore, the muons being relatively heavy particles, their emission is comparably reduced. The hard photons emitted from the electrons are cut-off from the geometrical restrictions for $\theta \sim 90°$, or, for small θ, they are cut-off again by the energy resolution. The integration is to be performed on the shadowed area of Fig. 12, i. e., all photons emitted with energy up to that corresponding to the contour of the shadowed area must be included.

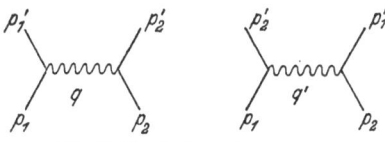

Fig. 13. Graphs for $e^- + e^- \to e^- + e^-$

As usual, one first integrates for an isotropic maximum photon energy, in order to conveniently eliminate the infrared divergencies, and then adds the remaining bremsstrahlung contribution. The "isotropic correction" (which includes also virtual photon corrections) is given by

$$\delta_R(\Delta E) = \frac{\alpha}{\pi}\left\{ -\frac{4}{3}\pi^2 - \frac{56}{9} + \frac{26}{3}\ln 2 + \frac{13}{3}\left(\ln\frac{E}{m_e} + \ln\frac{E}{m_\mu}\right) + \right.$$
$$\left. - 4\left(\ln\frac{E}{m_e} + \ln\frac{E}{m_\mu} + 2\ln 2 - 1\right)\ln\frac{E}{\Delta E}\right\}$$

(23)

where ΔE is the isotropic cut-off on the photon energy. The complete integration on the shadowed area has been carried out by G. LONGHI and has given the results:

E (MeV)	δ_R (90°)	δ_R (60°)	δ_R (30°)
200	-19%	-20%	-21%
400	-22%	-23%	-24%
600	-25%	-25%	-26%
800	-26%	-27%	-28%
1200	-28%	-29%	-30%
1500	-30%	-30%	-31%

One expects that, by relaxing the energy resolution, the radiative corrections may decrease in absolute value. The decrease is not however large for θ, say, between 60° and 90° essentially because one is now practically including the hard photons from the muons, which are already down because of the large muon mass.

The problem of higher order corrections is mathematically very complex. An attempt was made to estimate higher order terms by using the method of renormalization group by FURLAN, GATTO and LONGHI [11].

6. Electron-electron scattering

$$e^- + e^- \to e^- + e^-$$

(24)

arises in lowest order from the graphs of Fig. 13. The momentum transfers are both spacelike

$$q^2 = 4E^2 \sin^2 \frac{\theta}{2} \tag{25}$$

$$q'^2 = 4E^2 \cos^2 \frac{\theta}{2} . \tag{26}$$

The Møller formula, modified by the introduction of a form factor F, is, with neglect of the electron mass,

$$\frac{d\sigma}{d(\cos\theta)} = \frac{\pi\alpha^2}{4E^2} \left\{ \frac{4 + (1 + \cos\theta)^2}{(1 - \cos\theta)^2} |F(q^2)|^2 + \frac{8}{\sin^2\theta} \mathrm{Re}\left[F^*(q^2) F(q'^2)\right] + \right.$$
$$\left. + \frac{4 + (1 - \cos\theta)^2}{(1 + \cos\theta)^2} |F(q'^2)|^2 \right\} . \tag{27}$$

The form factor F can be thought of as a product $V_e^2 P_\gamma$, where V_e is the electron-vertex form factor and P_γ is the photon-propagator form factor. For $F = 1$ Eq. (27) takes the simple form

$$\frac{d\sigma}{d(\cos\theta)} = \frac{\pi\alpha^2}{2E^2} \left(\frac{3 + \cos^2\theta}{\sin^2\theta} \right)^2 . \tag{28}$$

Measurements of Møller scattering can be used to probe the theory. If we assume for F

$$F(q^2) = \frac{1}{1 + \dfrac{q^2}{Q^2}}$$

the relative deviation from the case $F = 1$ is the same as for electron-positron scattering. The maximum deviation again occurs at 90° as for electron-positron scattering. It is important for the interpretation of the experiment to know accurately the radiative corrections. Calculations have been done by Tsai [14] and by Bayer and Kheifets [15] under very similar assumptions. These calculations are made by properly taking into account the experimental conditions for the particular colliding beam experiment, and in this respect they differ from other calculations made in view of different (and sometimes irrealizable) conditions. The experiment considered by Tsai (and performed by the Stanford group) is based on the idealized arrangement of Fig. 14. The two detectors for the final electrons are in coincidence. They are centered around the angle θ. Their opening angle is $2\Delta\theta$. They are assumed to have no energy resolution. If a photon is emitted in the direction θ it can take an energy up to the kinematically allowed maximum energy $\sim E$, without being detected. In fact the geometric limits from the detection system do not impose any condition for photons emitted in a direction parallel to the final electrons. However a photon emitted in a direction parallel to the initial particles can only have energy up to ΔE

$$\Delta E = \frac{2E\,\Delta\theta}{\sin\theta + \Delta\theta} \tag{29}$$

consistently with the geometric conditions. The condition (29) is obtained directly from kinematics by considering the configuration for which the photon energy is maximum compatibly with the condition that the final electron momenta stay inside the two cones subtended by the detectors. For photons emitted along other directions the maximum energy allowed by the apparatus varies with angle. The kind of ellipsoid reported in

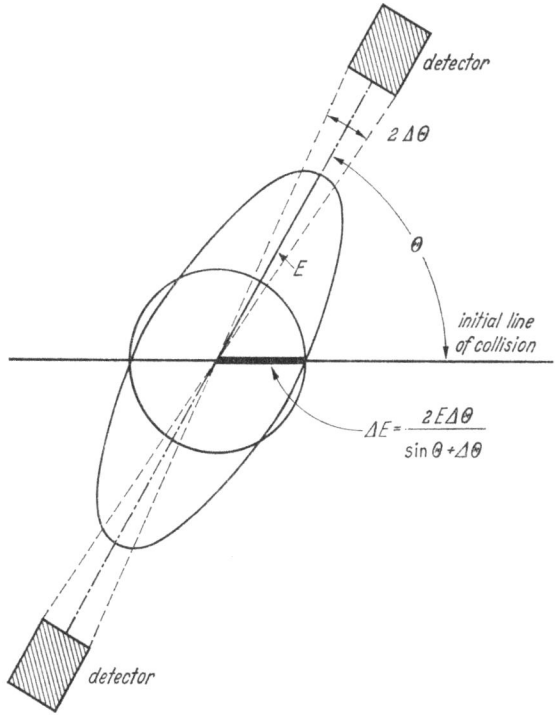

Fig. 14. Energy upper limits for a bremsstrahlung photon in the $e^- - e^-$ scattering experiment

Fig. 14 shows graphically the dependence of the maximum photon energy with photon direction. TSAI's calculations are done by first integrating up to $K_0 = \Delta E$ isotropically. The circle of radius ΔE is reported in Fig. 14. The addition of the "soft" photons with $K_0 < \Delta E$ to the virtual photon radiative corrections removes, as well-known, the dependence on the fictitious photon mass. To these "soft" corrections one then adds the hard photon corrections:

$$d\sigma = d\sigma_{el} + d\sigma_{soft} + d\sigma_{hard} . \tag{30}$$

The soft photon correction is well-known. One has

$$\left[\frac{d\sigma}{d(\cos\theta)}\right]_{el} + \left[\frac{d\sigma}{d(\cos\theta)}\right]_{soft} \cong \frac{\pi\alpha^2}{4E^2}\left[\frac{4 + (1+\cos\theta)^2}{(1-\cos\theta)^2} + \frac{4}{\sin^2\theta}\right] \times$$

$$\times \left\{1 - \frac{4\alpha}{\pi}\left[\frac{23}{18} - \frac{11}{12}\ln\left(\frac{2E^2}{m_e^2}(1-\cos\theta)\right)\right] + \right.$$

$$\left. + \left[\ln\left(\frac{E^2}{m_e^2}\sin^2\theta\right) - 1\right]\ln\frac{E}{\Delta E}\right\} + \text{same term with } \cos\theta \to -\cos\theta . \tag{31}$$

The integration of the hard portion is much more difficult and has been done by TSAI. For $E = 500$ MeV and $\Delta\theta = 3.5°$ TSAI finds a total correction [defined from $d\sigma = d\sigma_{\text{Møller}}(1 + \delta)$] of -9.5% at 90°, and of -6.0% at 35%. The calculations performed by BAYER and KHEIFETS are claimed to be more accurate. These authors introduce also a threshold resolution E_1 for the detectors: i. e. the detectors do not register electrons

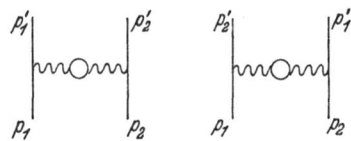

Fig. 15. Vaccum polarization graphs

with energy lower than E_1. Their numerical results are as follows. For $E = 100$ MeV, $E_1 = 50$ MeV and $\Delta\theta = 1.25°$ one has: $\delta(40°) = -15.8\%$, $\delta(50°) = -17.4\%$, $\delta(60°) = -18.3\%$, $\delta(70°) = -18.9\%$, $\delta(80°) = -19.4\%$, $\delta(90°) = -19.6\%$. For $E = 500$ MeV, $E_1 = 10$ MeV, and $\Delta\theta = 3.5°$: $\delta(40°) = -10.6\%$, $\delta(50°) = -12.1\%$, $\delta(60°) = -13.2\%$, $\delta(70°) = -13.7\%$, $\delta(80°) = -14.1\%$, and $\delta(90°) = -14.4\%$. The last choice of the parameters E and $\Delta\theta$ coincides with that of Tsai except for the threshold resolution $E_1 = 10$ MeV. The absolute values for $\delta(\theta)$ obtained by BAYER and KHEIFETS are however larger.

It is interesting to know the relative magnitude of the various contributions from virtual electrons, muons, pion and proton pairs in the vacuum polarization diagrams that contribute to the elastic radiative corrections (Fig. 15). Using known expressions by FEYNMAN [16] and by EUWEMA and WHEELER [17], valid for particles without structure, one sees that in $e^- + e^- \to e^- + e^-$ at $E = 500$ MeV and $\theta = 90°$ (where the contribution is expected to be larger) the vacuum polarization contribution is of 2% for electrons, 0.35% for muons, 0.089% for pions, and 0.02% for protons in the closed loops of Fig. 15.

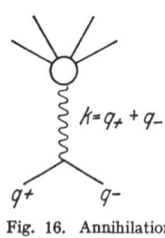

Fig. 16. Annihilation graph

7. The rest of this discussion — except for a few remarks at the end about weak interactions — will concern $e^+ - e^-$ annihilation into strong-interacting-particles. The reactions proceed mostly through one-photon exchange (Fig. 16).

More generally we shall talk of the one-photon channel whenever the interaction passes through one intermediate photon. For experiments that treat the charges symmetrically (such as total cross-sections or, for instance, $e^+ + e^- \to \pi^+ + \pi^-$ at a definite angle but without distinguishing the final charges) the interferences between the one-photon and the two-photon channels vanish on account of charge conjugation invariance. A simple proof is as follows. The interference term between one-photon and two-photon channel has the form

$$\text{Re} \sum_f \langle i | S_1 | f \rangle \langle f | S_2 | i \rangle = \text{Re} \langle i | S_1 P_f S_2 | i \rangle \qquad (32)$$

if S_1 is that part of the S matrix from the one-photon channel and S_2 the part from the two-photon channel. We have called P_f the projector into the final states f. We now assume that the set of states is symmetric under charge conjugation such that

$$C^{-1} P_f C = P_f$$

where C is the charge conjugation operator. Eq. (32) can be written as

$$(32) = \operatorname{Re} \langle i_- | S_1 P_f S_2 | i_+ \rangle = \operatorname{Re} \langle i_- | C^{-1} S_1 P_f S_2 C | i_+ \rangle = 0$$

where $|i_+\rangle$ and $|i_-\rangle$ are respectively even and odd under charge conjugation and we have noted that $S_1 |i\rangle = S_1 |i_-\rangle$ and $S_2 |i\rangle = S_2 |i_+\rangle$. For processes that go through the one-photon channel the final strong-interacting-particles produced according to

$$e^+ + e^- \to a + b + c + \cdots$$

must be in a state with $P = -1$, $C = -1$, $J = 1$, $T = 1$ or 0. That the final amplitude must transform as a polar vector follows at once by observing that the gauge condition $K_\mu \langle f | j_\mu | 0 \rangle = 0$ on the matrix element of the current j_μ, becomes in c. m., where $K_\mu = (0, K_4)$, $K_4 \langle f | j_4 | 0 \rangle = 0$ implying that the amplitude is given by the polar vector $\langle f | \vec{j} | 0 \rangle$. The one-photon channel selection rules, $P = -1$, $C = -1$, $J = 1$, $T = 1, 0$, are certainly valid to order $e^4 = \alpha^2$ in the intensity, and they are violated from interference with the two-photon channels at order $e^6 = \alpha^3$. However for the symmetrical experiments we have mentioned these interferences vanish and the violation of the selection rules only occurs at order $e^8 = \alpha^4$ in the intensity (one has to add brems-strahlung corrections). For production of n pions one has to add to the above rules the rule $T = 1$ for n even and $T = 0$ for n odd. A typical feature of two-body annihilation through the one-photon channel is the

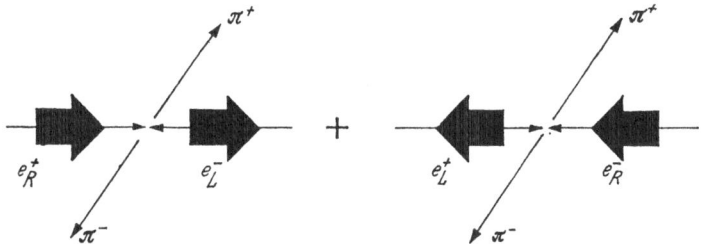

Fig. 17. Initial $e^+ e^-$ state

possibility of predicting at any energy the form of the angular distribution in the form $A + B \cos^2\theta$. For annihilation into two spin zero particles (2π, $\eta + \pi$, $2\,\mathrm{K}$, etc.) the final pions cannot move in the same direction of the incident pair if angular momentum has to be conserved (we are always neglecting the electron mass). In fact in the $m_e = 0$ limit a right-handed electron couples to a left-handed positron and a left-handed electron to a right-handed positron. If the pions were emitted along the collision line the angular momentum projection on the line would be zero in contrast to the initial value of $+1$ or -1 (see Fig. 17). Therefore

$A + B = 1$ and the distribution is proportional to $\sin^2\theta$. Two spinless bosons produced through the one-photon-channel will be in a P-state if their relative parity is positive. Two spin-one bosons of positive relative parity are produced in $1P_1$, $5P_1$ and $5F_1$. A pair of spin $\frac{1}{2}$ particles is produced in $3S_1$ and $3D_2$ for positive relative parity or in $1P_1$ and $3P_1$ if the relative parity is negative. Similarly a pair of a vector meson and a pseudoscalar meson will be produced in relative P state. In the one-photon channel the unitarity limitation on the cross-section takes a special simple form. From the limitation $\sigma_J \leq \pi \, \lambda^2 (2J + 1)$, which for $J = 1$ gives $\sigma_1 \leq 3\pi \, \lambda^2$ we get here the limitation

$$\sigma \leq \frac{3}{4} \pi \, \lambda^2 \tag{33}$$

for all reactions going through the one-photon channel. The factor $\frac{1}{4}$ comes from the fact that, in the limit $m_e = 0$ only one of the four possible initial $e^+ e^-$ spin states can contribute to the cross-section, namely the state

$$\frac{1}{\sqrt{2}} \left(|e_L^-, e_R^+\rangle + |e_R^-, e_L^+\rangle \right). \tag{34}$$

8. The reaction $e^+ + e^- \to a + b + \cdots$ is especially useful to measure the form factors of the vertex $(\gamma,\, a + b + \cdots)$ of the strong-interacting-particles a, b, ... for timelike value $K^2 = -4E^2$ of the virtual photon momentum. One important experimental aspect is the unique possibility

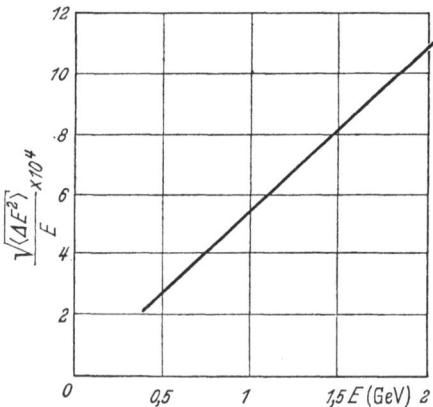

Fig. 18. Predicted energy dispersion of the beam in the Adone project

of disposing of very narrow energy resolutions which will allow measurements of very narrow widths of possible resonances in the cross-sections. The expected energy dispersion of the beam is illustrated in Fig. 18 (I would like to thank F. Amman and C. Bernardini for kindly providing me with this figure). These resonances are expected to occur when an intermediate boson of $P = -1$, $C = -1$, $J = 1$ and zero strangeness of mass $M^2 = 4E^2$ couples to the initial photon and to the final strong-interacting-particles. Intermediate bosons of different quantum numbers are expected not to contribute as they do not couple in the dominant one-photon channel. In more detail for a qualitative discussion of the possible resonant contributions we can assume a Breit-Wigner description for the resonance cross-section near its maximum

$$\sigma(E) \cong \pi \, \lambda^2 \frac{2J+1}{4} \frac{\Gamma_i \Gamma_f}{(2E-M)^2 + \dfrac{\Gamma^2}{4}} \tag{35}$$

where Γ_i and Γ_f are the rates into the initial $e^+ - e^-$ state and the final state respectively. If the energy resolution in c. m. is $2\Delta E$ what one observes is

$$\bar{\sigma}_R = \frac{1}{\Delta E} \int\limits_{\frac{1}{2}(M-\Delta E)}^{\frac{1}{2}(M+\Delta E)} \sigma(E) \, dE . \tag{36}$$

If the width is larger than $2\Delta E$, $\Gamma \gg 2\Delta E$, as it will most often occur, at the resonance one will observe

$$\sigma_R = \sigma(2M) = \pi \, \lambda^2 (2J+1) \, B_i \, B_f \tag{37}$$

where B_i and B_f are the branching ratios for decay into the initial and the final state respectively. If there is a dominant final state, $B_f \cong 1$ and the relevant quantity is

$$\frac{\Gamma_i}{\Gamma} . \tag{38}$$

The other situation, $\Gamma \ll 2\Delta E$, gives for the observed value

$$\bar{\sigma}_R \cong 2\pi \, \lambda^2 \frac{\pi}{4} (2J+1) \, B_i \, B_f \frac{\Gamma}{2\Delta E} . \tag{39}$$

Again, with $B_f \cong 1$, the relevant quantity is

$$\frac{\Gamma_i}{\Delta E} . \tag{40}$$

The possibility of observing the resonant contribution thus depends on the value of Γ_i/Γ for a wide (as compared to ΔE) resonance, or on the value of $\Gamma_i/\Delta E$ for a narrow resonance. If $\Gamma \sim \Delta E$ the two parameters coincide and the discussion can be extended also to these cases. Now Γ_i is strongly affected by some factors that can be derived from general considerations. If the resonant state has

$$
\begin{aligned}
J &= 0, \; C = \; 1 \quad \text{then} \quad && \Gamma_i \propto \alpha^4 \, m_e^2 \\
J &= 0, \; C = -1 \quad && \Gamma_i \propto \alpha^6 \, m_e^2 \\
J &= 1, \; C = \; 1 \quad && \Gamma_i \propto \alpha^4 \\
J &= 1, \; C = -1 \quad && \Gamma_i \propto \alpha^2 \\
J &= 2, \; C = \; 1 \quad && \Gamma_i \propto \alpha^4 \\
J &= 2, \; C = -1 \quad && \Gamma_i \propto \alpha^6 \\
&\qquad\qquad\qquad \text{etc.}
\end{aligned}
$$

These rules can easily be obtained from charge conjugation, angular momentum, and the helicity rule that we have already used before. The biggest effects are for $J = 1$, $C = -1$ (ρ^0, ω^0, φ^0) and next in importance for $J = 1$, $C = 1$ (perhaps A_1^0 (1080) has these quantum numbers; such states however are not visible in 2π annihilation) and $J = 2$, $C = 1$ (perhaps f^0 or A_2^0 (1310)), while for $J = 0$, $C = 1$ (π^0, η^0) there is no hope of an observable effect.

9. Annihilation into n pions through the one-photon channel leads, as we have said, to a final state with $P = -1$, $C = -1$, $J = 1$, $T = 1$ for n even and $T = 0$ for n odd. If $\vec{J}$ is the most general vector (pseudovector) formed out of the independent final momenta for n even (odd) the final distribution can be shown to have the general form

$$\frac{1}{2} |\vec{J}|^2 \sin^2\theta \qquad (41)$$

where θ is the angle formed between $\vec{J}$ and the initial line of collision. For annihilation into 2π the only vector is $\vec{p}_1 - \vec{p}_2$ and we get the expected $\sin^2\theta$ distribution. For annihilation into 3π the only pseudovector is the normal to the production plane, and we thus obtain that such a normal has a $\sin^2\theta$ distribution around the initial line of collision. For annihilation (by one-photon exchange) into two spin-less bosons the differential cross-section in c. m. is

$$\frac{d\sigma}{d(\cos\theta)} = \frac{\pi}{16} \alpha^2 \frac{1}{E^2} \beta^3 |F(K^2)|^2 \sin^2\theta \qquad (42)$$

where $F(K^2)$ is the e. m. form factor of the boson. The total cross-section is

$$\sigma = \frac{1}{m^2} (0.53 \cdot 10^{-32} \text{ cm}^2) \frac{1}{x^2} \left(1 - \frac{1}{x^2}\right)^{\frac{3}{2}} |F(-4m^2 x^2)|^2 \qquad (43)$$

where m is the mass of the boson in GeV and $x = E/m$. Eq. (42) applies to $e^+ + e^- \to \pi^+ + \pi^-$, $e^+ + e^- \to K^+ + K^-$, $e^+ + e^- \to K_1^0 + K_2^0$, etc. The reaction $e^+ + e^- \to \pi^0 + \pi^0$ does not proceed through the one-photon channel. It requires exchange of two photons. The $K^0 \bar{K}^0$ pair produced through one-photon channel must be in a state with $C = -1$. Therefore only $K_1^0 K_2^0$ pairs are produced, but no $K_2 K_2$ or $K_1 K_1$ pairs. From (42) or (43) one can measure the form factor $F(K^2)$ for $K^2 = -4E^2$, starting from the threshold $K^2 = -(2m_\pi)^2$. The form factor is complex in the observed physical region; however its phase cannot be directly determined from the experiment. The annihilation $e^+ + e^- \to \pi^+ + \pi^- + \pi^0$ leads to a final state with $l = L = 1, 3, 5$, etc., where l is the $\pi^+ - \pi^-$ relative angular momentum and L is the π^0 angular momentum relative to $\pi^+ - \pi^-$. One has

$$\frac{d^2\sigma}{d\omega_+ \, d\omega_- \, d(\cos\theta)} = \frac{\alpha}{(2\pi)^2} \frac{1}{64 E^2} |H|^2 \sin^2\theta \, (\vec{p}^{(+)} \times \vec{p}^{(-)})^2 \qquad (44)$$

where ω_+, ω_- are the energies of π^+, π^- in c. m. and $\vec{p}^{(+)}$, $\vec{p}^{(-)}$ are their momenta; H is a form factor depending on E, ω_+, ω_-, and θ, already defined, is the angle of the normal to the production plane with the initial line of collision. The dependence of H on its variables will be affected strongly from the existence of the ω meson and of the ρ meson. The ω-meson will be responsible for the expected resonant behaviour for $K^2 = -m_\omega^2$. The possibility of resonant interaction of the final mesons in the $J = 1, T = 1$ meson resonant state will appreciably affect the final distribution. From such considerations one would suggest to try first with H of the form

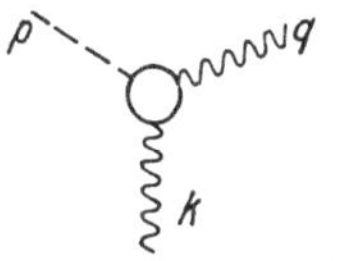

$$H = \frac{a}{4E^2 - m_\omega^2 + i\,m_\omega\,\Gamma_\omega} \times$$

$$\sum_i \times \frac{1}{(4E^2 + m_\pi^2 - m_\rho^2 + i\,m_\rho\,\Gamma_\rho)^2 - 4E\,\omega_i} \tag{45}$$

Fig. 19. The $\pi^0 - 2\gamma$ vertex

where ω_i is ω_+, ω_-, ω_0 for $i = 1, 2, 3$, Γ_ω, Γ_ρ and m_ω, m_ρ are the widths and masses of ω and ρ, and a at a first approximation could be assumed to be roughly constant. The mode $e^+ + e^- \to 3\pi^0$, or in general $\to n\,\pi^0$, is forbidden in the one-photon channel. The annihilation $e^- + e^- \to \pi^0 + \gamma$ through one-photon exchange, would allow the exploration of the $\pi^0 \to 2\gamma$ vertex. This vertex has the form

$$\frac{1}{4}\,G(-K^2, -q^2, -p^2)\,\varepsilon_{\mu\nu\lambda\varrho}\,F_{\mu\nu}(q)\,F_{\lambda\varrho}(K) \tag{46}$$

where G is a form factor depending on $K^2 = -4E^2$, the pion momentum p, and the photon momentum q (see Fig. 19). We write $G(-K^2)$ for $G(-K^2, 0, m_\pi^2)$. The π^0 lifetime is given by

$$\frac{1}{\tau} = \frac{m\pi^3}{64\,\pi}\,|G(0)|^2 \tag{47}$$

and the cross-section for $e^+ + e^- \to \pi^0 + \gamma$ is

$$\frac{d\sigma}{d(\cos\theta)} = \frac{\pi\,\alpha}{m_\pi^3}\,\frac{1}{\tau}\,\beta^3\,(1 + \cos^2\theta)\,\left|\frac{G(-K^2)}{G(0)}\right|^2. \tag{48}$$

Eq. (48) also applies, with the suitable changes, to $e^+ + e^- \to \eta^0 + \gamma$. Annihilation into a fermion-antifermion pair, according to

$$e^+ + e^- \to p + \bar{p}, \quad n + \bar{n}$$
$$\to \Lambda + \bar{\Lambda} \tag{49}$$
$$\to \Sigma + \bar{\Sigma}$$
$$\to \Xi + \bar{\Xi}$$

leads to a final state consisting of $^3S_1$ and $^3D_1$ as long as the process proceeds through the one-photon channel. Near threshold the cross-section will thus be isotropic and will rise $\propto \beta$. The differential cross-section is in general

$$\frac{d\sigma}{d(\cos\sigma)} = \frac{\pi}{8}\,\alpha^2\,\lambda^2\,\beta\,\left[|F_1(K^2)|^2 + \mu\,|F_2(K^2)|^2\,(1 + \cos^2\theta) + \right.$$
$$\left. + \left|\frac{m}{E}\,F_1(K^2) + \frac{E}{m}\,\mu\,F_2(K^2)\right|^2 \sin^2\theta\right] \tag{50}$$

where F_1 and F_2 are the Dirac and Pauli form factors respectively normalized to $F_1(0) = 1$ and $F_2(0) = 1$ for a charged fermion and μ is the static anomalous magnetic moment of the fermion. Along the absorptive cut the form factors are in general complex. The absorptive cut starts at much lower values of K^2 than the physical region for the annihilation. For instance in $e^+ + e^- \to N + \bar{N}$ the absorptive cut starts at $K^2 = -4 \mu_\pi^2$ for the isovector part and at $K^2 = -9 \mu_\pi^2$ for the isoscalar part. The fact that the form factors are in general complex in the physical region of the process brings about the existence of a polarization of the final fermions normal to the production plane, already at the lowest electromagnetic order. This situation is opposite to scattering $e +$ (fermion) $\to e +$ (fermion) where there is no polarization at lowest order. In $e^+ + e^- \to$ (fermion) $+$ (antifermion) the polarization of the fermion is given by $P(\theta)$, where

$$\frac{d\sigma}{d(\cos\theta)} P(\theta) = -\frac{\pi}{8} \alpha^2 \lambdabar^2 \beta^3 \mu \frac{E}{m} \, \mathrm{Im}\,[F_2^*(K^2)\, F_1(K^2)] \sin(2\theta) \qquad (51)$$

and it is directed along $\vec{p} \times \vec{q}_+$ where $\vec{p}$ is the momentum of the final fermion and $\vec{q}_+$ that of the positron. The polarization of the antifermion is $- P(\theta)$. The reactions $e^+ + e^- \to \Lambda^0 + \bar{\Sigma}^0$ and $e^+ + e^- \to \Sigma^0 + \bar{\Lambda}^0$ have a cross-section

$$\frac{d\sigma}{d(\cos\theta)} = \frac{\pi}{8} \alpha^2 \lambdabar^2 \beta \Big\{ \beta^2 \cos^2\theta \,[|f_1(K^2)|^2 + K^2 |f_2(K^2)|^2] +$$

$$[|f_1(K^2)|^2 - K^2 |f_2(K^2)|^2] \frac{E_\Lambda E_\Sigma + m_\Lambda m_\Sigma}{E^2} - 4 \, \frac{m_\Lambda E_\Sigma + m_\Sigma E_\Lambda}{E} \times$$

$$\times \mathrm{Re}\,[f_1(K^2)\, f_2^*(K^2)] \Big\} . \qquad (52)$$

The $\gamma \Sigma \Lambda$ vertex is written as

$$\bar{u}_\Lambda\,[f_1(K^2)\, \gamma_v + f_2(K^2)\, \sigma_{v\mu} K_\mu + f_3(K^2)\, K_v]\, v_\Sigma \qquad (53)$$

where u_Λ and v_Σ are the spinors and the f's are the form factors. The gauge condition implies in general

$$f_3(K^2) = i \frac{f_1(K^2)}{K^2} (m_\Sigma - m_\Lambda) . \qquad (54)$$

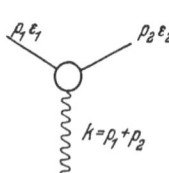

Fig. 20. The electromagnetic vertex of vector mesons

[However f_3 does not enter in (52). This can be seen directly from (53) in c. m. where $\vec{K} = 0$ and by noting that the fourth component of the current vanishes in c. m. from the condition $K_v j_v = K_4 j_4 = 0$.]

9. Production of a pair of spin one bosons

$$e^+ + e^- \to B + \bar{B}$$

occurs in the states $^1P_1$, $^5P_1$, and $^5F_1$ as can be seen from angular momentum, parity and charge conjugation. The $\gamma B B$ vertex is shown in Fig. 20. The momenta p_1 and p_2 satisfy $p_1^2 = p_2^2 = -m_B^2$ and $(p_1 \varepsilon_1) = (p_2 \varepsilon_2) = 0$.

The general form of the current is

$$
J_\mu = \frac{e}{(4\omega_1\omega_2)^{1/2}} \Big\{ G_1(\varepsilon_1\,\varepsilon_2)\,p_\mu + [G_1 + \mu G_2 + \varepsilon G_3] \times
$$
$$
\times [(\varepsilon_1\,K)\,\varepsilon_{2\mu} - (\varepsilon_2\,K)\,\varepsilon_{1\mu}] +
$$
$$
+ \varepsilon\,\frac{1}{m_B^2}\,G_3[(K\,\varepsilon_1)\,(K\,\varepsilon_2) - \frac{1}{2}\,K^2(\varepsilon_1\,\varepsilon_2)]\,p_\mu \Big\} \tag{55}
$$

where ω_1 and ω_2 are the c. m. energies of B, $\overline{\text{B}}$; $e\,G_1$, $\mu\,G_2$, $\varepsilon\,G_3$ describe the charge, the magnetic moment and the electric quadrupole moment distributions of B ($\mu + \varepsilon$ and 2ε are the static magnetic moment and electric quadrupole moment respectively). The G's are functions of K^2. The cross-section is

$$
\frac{d\sigma}{d(\cos\theta)} = \frac{\pi}{16}\,\alpha^2\,\lambda^2\,\beta^3\,\Big\{ 2\left(\frac{E}{m_B}\right)^2 |G_1 + \mu G_2 + \varepsilon G_3|^2(1 + \cos^2\theta) +
$$
$$
+ \sin^2\theta\,\Big[2\Big| G_1 + 2\left(\frac{E}{m_B}\right)^2 \varepsilon G_3 \Big|^2 + \Big| G_1 + 2\left(\frac{E}{m_B}\right)^2 \mu G_2 \Big|^2 \Big] \Big\}. \tag{56}
$$

The β^3 dependence (β is the final velocity) is typical of P-state production. With $G_1 = 1$, $G_2 = G_3 = 0$ the total cross-section is

$$
\sigma = m_B^{-2}\,(2.1 \cdot 10^{-32}\ \text{cm}^2)\,\frac{3}{4}\,(1 - u)^{\frac{3}{2}}\left(\frac{4}{3} + u\right) \tag{57}
$$

where m_B is in MeV and u is $(m/E)^2$. The cross-section (57) violates unitarity at high energies. In fact it goes to a constant value when $E \to \infty$ (i. e. $u \to 0$). On the other hand our unitarity limitation, Eq. (33), for the one-photon channel decreases with energy proportionally to $\lambda^2 = E^{-2}$. Although in general one would expect that the form factors in (56) [which have been ignored in (57)] will produce the required damping much before the unitarity limit is reached or the one-photon-channel approximation is unapplicable, the production of weak-inter-acting-vector mesons might with a good approximation be represented by the cross-section (57). In fact such bosons are not supposed to have strong interactions and their interaction with photons could well be approximated with that of point charges up to rather high energies. The violation of unitarity, by comparing Eq. (33) with Eq. (57) would in any case occur only at $E \sim 100\ m_B$. The total cross-section, from Eq. (57), is given in Fig. 21 for various values of m_B as a function of $x = E/m_B$. One sees that if one could dispose of colliding $e^+ - e^-$ beams of sufficiently high energy it would not be difficult to improve the present lower limit on the mass of the intermediate vector mesons.

10. Experimental investigations about the limits of validity of quantum electrodynamics may eventually run into the difficulty of having to deal with effects of the same order as the effects originating from virtual strong interacting particles. These effects are hard to estimate accurately. It may be relevant to note that direct measurements of the production of strong interacting particles in $e^+ - e^-$ collisions can directly be related to quantities that express the effect of virtual strong

interacting particles on the electrodynamic parameters. The modification
to the photon propagator can be expressed through a function $\pi(K^2)$,

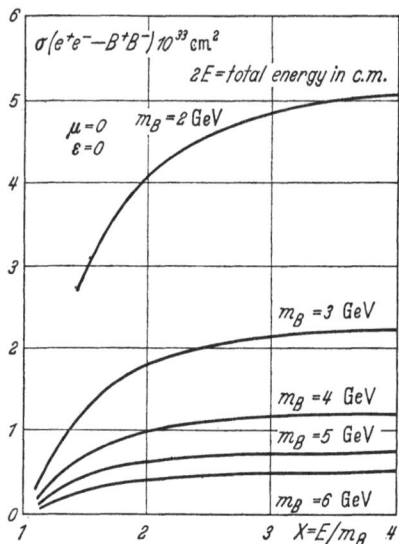

Fig. 21. Total cross-section for annihilation into a vector meson pair

additive with respect to the contributions from the various virtual
states [18]. The modified photon propagator can be written as

$$D'_{\mu v}(K) = \delta_{\mu v}\frac{1}{K^2} + \left(\delta_{\mu v} - \frac{K_\mu K_v}{K^2}\right) \int_0^\infty \frac{da\,\pi(-a)}{a(K^2+a-i\,\varepsilon)} \tag{58}$$

where

$$\pi(K^2) = -\frac{(2\pi)^3}{3K^2}\sum_{\substack{Z\\ p_Z=K}} \langle 0|j_v(0)|Z\rangle\langle Z|j_v(0)|0\rangle. \tag{59}$$

In (59) j_v is the current operator and the sum is extended over all physical
states with momentum $p^{(Z)}= K$. We consider the annihilation of $e^+ + e^-$
into a set of states Z and call $\sigma_Z(E)$ the total cross-section for such
annihilations. Similarly we call $\pi_Z(-4E^2)$ the contribution to (59)
from the set of states Z. One can formally show that

$$\pi_Z(-4E^2) = \frac{E^2}{\pi^2\,\alpha}\,\sigma_Z(E). \tag{60}$$

Eq. (60) is obtained by direct comparison of the two expressions (59)
and the usual expression for $\sigma_Z(E)$. The amplitude for $e^+ - e^-$ annihilation into a final state f, including lowest order vacuum polarization
effects due to strong interacting particles, can be written as

$$A(e^+ e^- \to f) = \left[1 + K^2\int_0^\infty \frac{da\,\pi(-a)}{a(K^2+a-i\,\varepsilon)}\right]\frac{2\pi e}{K^2}\,(\bar v\,\gamma_v u)\,\langle f|j_v(0)|0\rangle \times$$
$$\times\,\delta(p^{(+)}+p^{(-)}-p^{(f)}). \tag{61}$$

The cross-section for $e^+ + e^- \to f$ is thus modified, by inclusion of vacuum polarization effects due to strong interacting particles at lowest electromagnetic order, by the multiplicative factor

$$\mathfrak{J}(E) = \left| 1 - 4E^2 \int_0^\infty \frac{da \, \pi(-a)}{a(-4E^2 + a - i\varepsilon)} \right|^2. \tag{62}$$

We are particularly interested in the contribution to (62) from strong resonant states with the appropriate quantum numbers to couple to $e^+ - e^-$ at lowest e. m. order. We call the mass of the resonance M, its full width Γ, and B its branching ratio for decay into $e^+ - e^-$. For a rough estimate of $\mathfrak{J}(E)$ we approximate the resonant cross-section in the neighbourhood of the resonance with a Breit-Wigner formula

$$\sigma_R(E) = \pi \, \lambdabar^2 \, \frac{3}{4} \, B \, \frac{\Gamma^2}{(2E - M)^2 + \frac{1}{4}\Gamma^2}. \tag{63}$$

From (62), (60), and (63) one can derive an approximate expression for the resonant contribution to $\mathfrak{J}(E)$

$$\mathfrak{J}_R(E) \cong \left| 1 - \frac{3}{2} \, \alpha \, b \, \frac{\Gamma}{M - 2E - i\frac{\Gamma}{2}} \right|^2 \tag{64}$$

where we have introduced $b = B/\alpha^2$ to exhibit clearly the electromagnetic order. For a narrow resonance such that $\Gamma \ll \Delta E$ the effect will have to be averaged over ΔE giving

$$\overline{d\sigma} = \left[1 + \frac{\Gamma}{\Delta E} \, (1 + 9\alpha^2 \, b^2) \, \frac{\pi}{4} \right] d\sigma_0 \tag{65}$$

where $d\sigma_0$ is the lowest order cross-section. For a sufficiently wide resonance one can hope to measure $\mathfrak{J}(E)$ as given by (64). It can be seen that $\mathfrak{J}(E)$ has a maximum at about $2E = M + \frac{1}{2}\Gamma$ and a minimum at about $2E = M - \frac{1}{2}\Gamma$ where it reaches the values

$$1 + 3\alpha b \quad \text{and} \quad 1 - 3\alpha b \tag{66}$$

respectively. For the ω-meson, assuming a branching ratio of $1.0 \cdot 10^{-4}$ [19] for the decay mode $\omega \to e^+ + e^-$ we find that $\mathfrak{J}_{max} - 1$ from the ω resonance is $\sim 0.5 \cdot 10^{-1}$, which is much larger than other non-resonant contributions to vacuum polarization, but still of the order of the radiative corrections for the usual experimental arrangements. For the ρ meson with $B \sim 0.5 \cdot 10^{-4}$ one finds again an effect of $4 \cdot 10^{-2}$. However for the ρ meson, with a branching ratio into lepton pairs of $\sim 1.0 \cdot 10^{-3}$ the effect is expected to be very large, of the order of 30%.

11. We shall now look more closely at the annihilation modes into 2π, 3π, $2K$, and $\pi^0 + \gamma$, going through the known meson resonances. I shall report on some computations done with G. ALTARELLI and E. CELEGHINI.

Let us first discuss the particularly simple case of $e^+ + e^- \to (\rho^0) \to \pi^+ + \pi^-$. We shall describe this process by the graph of Fig. 22.

Fig. 22. The ρ^0 resonant contribution to $e^- + e^+ \to \pi^+ + \pi^-$

For its evaluation we need the $\gamma_{\rho\gamma}$ coupling constant of ρ^0 to a photon and the γ_ρ coupling constant of ρ^0 to $\pi^+ + \pi^-$. The two couplings are related as was shown by GELL-MANN and ZACHARIASEN [20] and by R. DASHEN and D. SHARP [21] by

$$\gamma_{\rho\gamma} = \frac{e \, m_\rho^2}{2\gamma_\rho}.$$

In general a similar situation is met when a meson (in this case ρ) is coupled to a current (in this case the isotopic spin current), which is in its turn coupled to the photon with the universal strength e. With such a model in the $e^+ + e^- \to (\rho^0) \to \pi^+ + \pi^-$ amplitude the coupling-constants $\gamma_{\rho\gamma}$ and γ_ρ cancel out and one has

$$\sigma = \frac{\pi \, \alpha^2}{12} \, \beta^3 \, \frac{1}{E^2} \, \frac{m_\rho^4}{[(4E^2 - m_\rho^2)^2 + m_\rho^2 \, \Gamma_\rho^2]}. \tag{67}$$

If calculated at resonance σ becomes

$$\sigma_R = \frac{\pi \, \alpha^2}{3} \, \beta^3 \, \frac{1}{\Gamma_\rho^2}. \tag{68}$$

This formula may be compared with what one gets from a Breit-Wigner formula at the resonance

$$\sigma_{R \, \text{(Breit-Wigner)}} = \frac{12\pi}{m_\rho^2} \, \frac{\Gamma_{e^+e^-}}{\Gamma_\rho}.$$

The proportionality in Eq. (68) to Γ_ρ^{-2} reflects the fact that $\Gamma_{e^+e^-}$ is inversely proportional to Γ_ρ: $\Gamma_{e^+e^-} \propto \alpha^2 m_\rho \left(\dfrac{m_\rho}{\Gamma_\rho}\right)$ such that $\sigma_{R\text{(Breit-Wigner)}}$ $\propto \dfrac{1}{m_\rho^2} \, \alpha^2 m_\rho \left(\dfrac{m_\rho}{\Gamma_\rho}\right) \dfrac{1}{\Gamma_\rho} \sim \dfrac{\alpha^2}{\Gamma_\rho^2}$. Such an inverse proportionality of $\Gamma_{e^+e^-}$ to Γ_ρ follows from the expression of $\gamma_{\rho\gamma}$. Note also that Eq. (67) can be derived also from a direct dispersion theoretic approach if one neglects 4π intermediate states and takes a Breit-Wigner shape for the resonating two-pion p-wave. Numerically with $\Gamma_\rho = 106$ MeV one gets from Eq. (68)

$$\sigma_R = 1.6 \cdot 10^{-30} \, \text{cm}^2 \quad (\text{with } \Gamma_\rho = 106 \, \text{MeV}).$$

Independently one can use directly the Breit-Wigner prediction and insert for $\Gamma_{e^+e^-}$, the width for $\rho \to e^+ e^-$ in the empirical number given by ZDANIS et al. [19]. With the quoted experimental errors one finds

$$\sigma_R \sim \text{between } (0.4 \text{ and } 2.5) \times 10^{-30} \, \text{cm}^2$$

not in contradiction with the above (completely theoretical) prediction. The same kind of argument can be applied to the reaction $e^+ + e^- \rightarrow$ $\rightarrow (\varphi^0) \rightarrow K^+ + K^+$ described by Fig. 23. At the $\varphi - \gamma$ vertex we insert $\gamma_{\varphi\gamma}$ and at the $\varphi \rightarrow K^+ K^-$ vertex we insert the coupling constant f_φ.

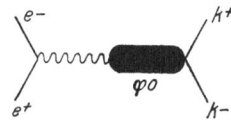

Fig. 23. The φ^0 resonant contribution to $e^+ + e^- \rightarrow K^+ + K^-$

The calculation of $\gamma_{\varphi\gamma}$ goes along the following line. One thinks of the physical φ as a superposition of the octet member φ_0 and the unitary singlet ω_0

$$\varphi = \cos\theta \; \varphi_0 + \sin\theta \; \omega_0 \,.$$

The e. m. current in SU_3 is supposed to have octet behaviour (except for special kinds of triplet models). This means that it cannot couple invariantly to φ_0; it is only coupled to φ_0. Now φ_0 is supposed to be coupled to the hypercharge current, with a strength that we call f_Y. Thus the coupling of φ_0 to the photon follows from the argument reported before for the case of ρ. If we then call $\cos\theta = a$ (and later on $\sin\theta = b$) we have

$$\gamma_{\gamma\varphi} = - a \, \frac{e \, m_\varphi^2}{2 f_Y}$$

(we have put m_φ^2 in the numerator for $m_{\varphi_0}^2$, without any appreciable error). The value f_Y can be obtained from f_ρ as φ_0 and ρ are members of the same vector meson octet. With such a model

$$\sigma = \frac{\pi \, \alpha^2}{12} \, a^2 \left(\frac{f_Y^2}{4\pi} \right)^{-1} \left(\frac{f_\varphi^2}{4\pi} \right) \frac{\beta^3}{E^2} \frac{m_\varphi^4}{[(4 E^2 - m_\varphi^2)^2 - m_\varphi^2 \, \Gamma_\varphi^2]} \,. \qquad (69)$$

For Γ_φ we assume $\Gamma_\varphi \cong 3.1$ MeV and we can obtain, from its value, f_φ^2 on the assumption that φ^0 decays mostly into $2\,\mathrm{K}$. At resonance we thus find, from (69),

$$\sigma_R = 0.5 \cdot 10^{-29} \, \mathrm{cm}^2 \,.$$

Similarly for $e^+ + e^- \rightarrow K^0 + \overline{K}{}^0$ we find $\sigma_R = 0.34 \cdot 10^{-29} \, \mathrm{cm}^2$. For $e^+ + e^- \rightarrow (\omega^0) \rightarrow \pi^+ \, \pi^- \, \pi^0$ the model is represented by Fig. 24.

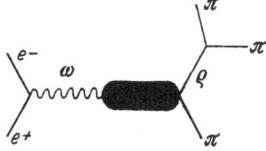

Fig. 24. The ω^0 resonant contribution to $e^+ + e^- \rightarrow 3\pi$

We need $\gamma_{\omega\gamma}$ that we can take, by the same argument used with φ, from the expansion $\omega = - \sin\theta \; \varphi_0 + \cos\theta \; \omega_0$ and from the $\varphi_0 - \gamma$

coupling

$$\gamma_{\omega\gamma} = b \frac{e \, m_\omega^2}{2 f_\gamma}.$$

We also need $f_{\omega\rho\pi}$ that is calculated (see for instance the paper by
DASHEN and SHARP) from the ω-width (taken ~ 9.5 MeV) assuming
that $\omega \to 3\pi$ goes as $\omega \to \rho + \pi \to (2\pi) + \pi$. Let us consider the particular configuration with the 3π emitted in $e^+ + e^- \to 3\pi$ of equal
energies in c. m. at 120° degrees with respect to each other and lying in a
plane containing the colliding beam direction. The choice of this particular configuration is suggested by the consideration of the $\sin^2\theta$ dependence of the normal to the final 3π plane with respect to the incident
momenta, which gives maximum probability for a final plane containing
the initial momenta. Furthermore the maximum probability with respect
to the final variables in the plane is obtained for an equilateral triangle
configuration, of the final pion momenta. For such a configuration at the
resonance

$$\frac{d\sigma}{d\omega_+ d\omega_-} = 0.7 \cdot 10^{-28} \text{ cm}^2/(\text{GeV})^2.$$

For the total cross-section at resonance we find $\sigma_R = 1.7 \cdot 10^{-30}$ cm^2.
We also note that with a Breit-Wigner expression and with the empirical
$\omega \to e^+ + e^-$ rate we obtain a total cross-section for $e^+ + e^- \to 3\pi$, at the
resonance, of $(0.4-4.7) \cdot 10^{-30}$ cm^2, where the uncertainty is that suggested by the given experimental error of the $\omega \to 2e$ rate of ZDANIS
et al. It is an interesting topic to study the effect of addition of two
photon exchange terms. Let us consider $e^+ + e^- \to \pi^+ + \pi^-$ and assume
that one photon exchange goes entirely through the intermediate ρ^0,
in the way we have already discussed. When considering two-photon
exchange, at order α^4, we expect pole contributions from $J = 1$, $C = 1$
mesons and $J = 2$, $C = 1$ mesons. However a $J = 1$, $C = 1$ meson cannot
decay into 2π (two pions in $l = 1$ have $C = -1$). Possible $J = 2$, $C = 1$
mesons are f^0 and A$_2$(1310). We shall discuss here the f^0 contribution.

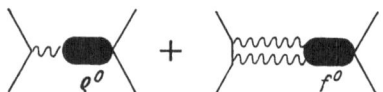

Fig. 25. Addition of two photons contribution to $e^+ + e^- \to 2\pi$

The model is shown in Fig. 25. We assume tentatively a direct f$^0 \to 2e$
coupling of the kind

$$\frac{i \, g}{8} \{ [\partial_\nu \bar{\psi}, \gamma_\mu \psi] - [\bar{\psi} \gamma_\mu, \partial_\nu \psi] + (\mu \to \nu)\} C^{\mu\nu}$$

where $C^{\mu\nu}$ describes f^0 and ψ the electron field. For g we make a guess:
$g \sim e^4/\pi^2 \, m_{\text{f}^0}$. The f$^0 \to 2\pi$ width is taken ~ 90 MeV. The general form of
the differential cross-section will be (independently of course of the
preceding special assumptions)

$$\frac{d\sigma}{d(\cos\theta)} \sim (A + B \cos\theta + C \cos^2\theta) \sin^2\theta.$$

We see that the $\sin^2\theta$ factor persists also with two photon exchange, in accordance with our helicity arguments (which are independent of the assumption of single γ exchange). The total cross-section for $e^+ + e^- \to$ $\to \pi^+ + \pi^-$ is now

$$\sigma = \frac{4}{3} A + \frac{4}{15} C$$

(no interference between 1γ channel and 2γ-channel for symmetric experiments) and C is also related to $\sigma_{2\pi^0}$ (which only goes through 2γ)

Fig. 26. Resonant contributions to $e^+ + e^- \to \pi^0 + \gamma$

by $\sigma_0 = \frac{2}{15} C$. On the special assumption for the $f^0 \to e^+ + e^-$ coupling reported above we find at the f^0 pole $A \simeq 3 \cdot 10^{-33}$ cm², $B \simeq 1 \cdot 10^{-33}$ cm², $C \simeq 1 \cdot 10^{-33}$ cm². The 2γ effect seems thus to be possibly of the same order as the 1γ effects (provided the assumption made of coupling only through a ρ^0 is verified). In any case the 2γ effect could perhaps be observable when the techniques will have reached a good stage of precision. Let us discuss next the $e^+ + e^- \to \pi^0 + \gamma$ reaction near the contributing poles ω^0, ρ^0, and φ^0. The diagrams are of the kind of Fig. 26. The couplings $\gamma_{\rho\gamma}$, $\gamma_{\omega\gamma}$, and $\gamma_{\varphi\gamma}$ are the same as before. We take the couplings $f_{\omega\pi\gamma}$, $f_{\rho\pi\gamma}$, and $f_{\varphi\pi\gamma}$ from the work of DASHEN and SHARP. The constant $f_{\omega\pi\gamma}$ is calculated assuming that the $\gamma \to \omega\pi$ transition goes through a ρ giving $f_{\omega\pi\gamma} \approx \gamma_{\rho\gamma} f_{\omega\rho\pi}/m_\rho^2 = e f_{\omega\rho\pi}/f_\rho (f_\rho = 2\gamma_\rho)$. Similarly $f_{\rho\pi\gamma}$ is assumed to go through a Y and one obtains $f_{\rho\pi\gamma}$ $= \gamma_{Y\gamma} \frac{1}{m_Y^2} f_{\rho Y\pi} = e(a f_{\varphi\rho\pi} - b f_{\omega\rho\pi})/2 f_Y$. Finally, by the same procedure, $f_{\varphi\pi\gamma} = e f_{\varphi\rho\pi}/f_\rho$. The coupling $f_{\omega\rho\pi}$ is obtained from the $\omega \to 3\pi$ width. The coupling $f_{\varphi\rho\pi}$ is not well known [from the $(\varphi \to \rho\pi)/(\varphi \to K\bar{K})$ branching ratio one obtains $f_{\omega\rho\pi}^2/f_{\varphi\rho\pi}^2 \sim 850$]. The model gives the following results. Around the ω resonance

$$\sigma = \frac{1}{24} \pi \alpha^2 b^2 \frac{f_{\omega\pi\gamma}^2}{f_Y^2} \frac{|\vec{p}|^3}{E^3} \frac{m_\omega^4}{(4E^2 - m_\omega^2)^2 + m_\omega^2 \Gamma_\omega^2} .$$

With $f_{\omega\pi\gamma} = e f_{\omega\rho\pi}/f_\rho = (9.2 \pm 0.9) \cdot 10^{-1}$ GeV⁻¹ and $\Gamma_\omega = 9.5$ MeV one obtains at resonance $\sigma_R = 3 \cdot 10^{-31}$ cm². For $\pi^0 + \gamma$ from the ρ^0 resonance one has similarly

$$\sigma = \frac{1}{6} \pi \alpha^2 \frac{f_{\rho\pi\gamma}^2}{f_\rho^2} \frac{|\vec{p}|^3}{E^3} \frac{m_\rho^4}{(4E^2 - m_\rho^2)^2 + m_\rho^2 \Gamma_\rho^2} .$$

With $f_{\rho\pi\gamma} = e f_{Y\rho\pi}/(2 f_Y) = 5.6 \cdot 10^{-1}$ GeV⁻¹ at resonance one has $\sigma = 7.5 \cdot 10^{-33}$ cm². Finally around the φ^0 resonance

$$\sigma = \frac{1}{24} \pi \alpha^2 a^2 \frac{f_{\varphi\pi\gamma}^2}{f_Y^2} \frac{|\vec{p}|^3}{E^3} \frac{m_\varphi^4}{[(4E^2 - m_\varphi^2)^2 + m_\varphi^2 \Gamma_\varphi^2]}$$

and with $f_{\varphi\pi\gamma} = e\, f_{\varphi\rho\pi}/f_\rho = (3.1 \pm 0.3) \cdot 10^{-2}\,\mathrm{GeV}^{-1}$ one finds $\sigma = 0.85 \cdot 10^{-31}\,\mathrm{cm}^2$.

12. I shall add some further remarks about weak interactions. We have already discussed the possible pair production of the intermediate bosons with colliding beams of sufficiently high energy. A check of the existence of a possible neutral semiweakly-interacting vector boson can be carried out with $e^+ - e^-$ beams of lower energy. If such a meson, B^0, exists and is coupled to neutral lepton currents it can show off as a resonance in

$$e^+ + e^- \to B^0 \to e^+ + e^-$$
$$e^+ + e^- \to B^0 \to \mu^+ + \mu^- .$$

Such resonances could lead to observable effects in spite of the fact that two semi-weak couplings are involved. The resonance being very narrow one will measure the average cross-section

$$\bar{\sigma}_R = \frac{3}{2}\, \pi^2\, \lambdabar^2\, B_i\, B_f\, \frac{\Gamma}{2\Delta E} .$$

Let us compare this cross-section to the (purely electrodynamic) cross-section $\sigma_0(2\mu)$ of $e^+ + e^- \to \mu^+ + \mu^-$ at high energy, which, for $E \gg m_\mu$, is $\sim \frac{1}{3}\, \pi\, \alpha^2\, \lambdabar^2$. We find

$$\frac{\bar{\sigma}_R(B^0)}{\sigma_0(2\mu)} = \frac{9}{2}\, \pi\, B_i\, B_f \left(\frac{\Gamma}{2\Delta E}\right) \frac{1}{\alpha^2}$$

where B_i and B_f are the branching ratios of B^0 into the initial and final state. Consider $e^+ + e^- \to \mu^+ + \mu^-$ and take $B_i \cong B_f \cong \frac{1}{5}$. A typical semiweak Γ corresponds to a few hundreds of eV. Take $\Gamma \sim 0.5 \cdot 10^{-3}$ MeV and $2\Delta E \sim 5$ MeV. One finds

$$\frac{\bar{\sigma}_R(B^0)}{\sigma_0(2\mu)} \cong 1$$

showing that the resulting effect would indeed be very large. From local weak interactions of typical weak strength we do not expect strong effects unless the colliding beams have an energy larger than ~ 10 BeV. For instance a weak lepton interaction of the form $(\mu^+ \mu^-)\,(e^+ e^-)$, if it exists, would give a coherent addition to the electromagnetic amplitude for $e^+ + e^- \to \mu^+ + \mu^-$. This small contribution will increase fastly with energy and bring about a $\cos\theta$ term in the angular distribution:

$$\frac{d\sigma}{d(\cos\theta)} \cong \frac{\pi}{8}\, \alpha^2\, \lambdabar^2\, [(1 + \cos^2\theta)\,(1 + \varepsilon) + 2\varepsilon \cos\theta]$$

where $\varepsilon \cong 6.2 \cdot 10^{-4}\,(E/M_N)^2$ (M_N = nucleon mass), if one assumes a weak coupling of the same strength of the β-coupling constant G. The $\cos\theta$ term would however also arise from interference of one-photon and two-photon exchange graphs. A typical parity-non-conserving effect like the longitudinal muon polarization would however be a clear-cut test for a weak interaction. The final $\mu^{\pm}$ longitudinal polarization is

$$P^{\pm} \cong \pm \varepsilon \frac{(1 + \cos\theta)^2}{1 + \cos^2\theta} .$$

Only for energies $E \sim 30$ GeV, E becomes of order unity and the polarization effect would be large.

In conclusion I would like to thank Doctors G. ALTARELLI, E. CELEGHINI and G. LONGHI for their precious collaboration to the preparation of these notes and to some of the calculations reported.

References

[1] DRELL, S. D.: Ann. phys. 4, 75 (1958)
[2] — and J. W. MCCLURE: To be published
[3] BAYER, V. N., and V. M. GALITSKY: Physics Letters 13, 355 (1964)
[4] YENNIE, D. R., S. C. FRAUTSCHI, and H. SUURA: Ann. phys. 13, 379 (1961)
[5] TSAI, Y. S.: Phys. Rev. 137, 730 (1965)
[6] ANDREASSI, G., G. CALUCCI, G. FURLAN, G. PERESSUTTI, and P. CAZZOLA: Phys. Rev. 128, 1425 (1962); see also ANDREASSI, G., P. BUDINI, and G. FURLAN: Phys. Rev. Letters 8, 184 (1962)
[7] ERIKSSON, K. E., and A. PETERMANN: Phys. Rev. Letters 5, 446 (1960) ERIKSSON, K. E.: Nuovo cimento 19, 1044 (1961)
[8] BERNARDINI, C., G. F. CORAZZA, G. DI GIUGNO, J. HAISSINSKI, P. MARIN, R. QUERZOLI, and B. TOUSCHEK: LNF — 64/33
[9] GARYBIAN, G. M.: Izvest. Akad. Nauk. Armenyan. S.S.R. 5, 3 (1952)
[10] ALTARELLI, G., and F. BUCCELLA: Nuovo cimento 34, 1337 (1964)
[11] FURLAN, G., R. GATTO, and G. LONGHI: Physics Letters 12, 262 (1964)
[12] MOSCO, U.: Nuovo cimento 33, 115 (1964)
[13] LONGHI, G.: Nuovo cimento 35, 1122 (1965)
[14] TSAI, Y. S.: Phys. Rev. 120, 269 (1960)
[15] BAYER, V. N., and S. A. KHEIFETS: Nuclear Phys. 47, 313 (1963)
[16] FEYNMAN, R. P.: Phys. Rev. 76, 769 (1949)
[17] EUWEMA, R., and J. A. WHEELER: Phys. Rev. 103, 803 (1965)
[18] KÄLLEN, G.: Quantenelektrodynamik, in Handbuch der Physik, Band V, Teil 1. Berlin-Göttingen-Heidelberg: Springer 1958
[19] ZDANIS, R. A., L. MADANSKY, R. W. KRAEMER, S. HERTZBACH, and R. STRAND: Phys. Rev. Letters 14, 721 (1965)
[20] GELL-MANN, M., and F. ZACHARIASEN: Phys. Rev. 124, 965 (1961)
[21] DASHEN, R., and D. SHARP: Phys. Rev. 133, 1585 (1964)

Prof. Dr. R. GATTO

Istituto di Fisica dell'Università, Firenze and
Laboratori Nazionali del C.N.E.N. di Frascati, Roma

Experimental Techniques*

W. K. H. Panofsky

Stanford Linear Accelerator,
Stanford University,
Stanford, California

In most general terms the techniques of experimental high energy physics are intermediate between two extremes: those methods in which essentially all events initiated in a primary of secondary target are registered in the apparatus, and those in which selectivity is involved before primary data are recorded. Approximating the first category are the track chambers which will form the subject of another talk at this conference. As to the items in the second category I shall restrict myself to two: (a) magnetic spectrometers, and (b) selection of photon energies It is perhaps appropriate that these two topics be emphasized in a conference devoted to high-energy electromagnetic interactions. In such interactions usually only a very small fraction of the events are low-momentum transfer, purely electromagnetic events which, at least in principle, do not contain new information. Hence "pre-selection" is of greater importance in this area in the other branches of high-energy physics.

Spectrometers used in high energy physics were generally developed initially by scaling and adapting the single lens, point-to-point focusing devices of low-energy nuclear physics. However, as energies enter the multi-BeV range, this scaling process loses validity. The reasons are: (1) Momentum resolution must be matched by angular resolution to correspond to a consistent value of center-of-mass energy resolution. Algebraically this requires an angular resolution δ_θ related to the fractional momentum resolution δ_p given by $\delta_\theta = \delta_p \left\{ \tan \dfrac{\theta}{2} + \dfrac{M}{E_1 \sin\theta} \right\}$, where E_1 is the incident energy, M is the rest energy of the target nucleon, and θ is the laboratory scattering angle; the first term dominates at high energies and intermediate angles. (2) Liquid hydrogen and deuterium targets are used for almost all experiments; therefore long target lengths must be used. As long as the target length is less than 0.02 to 0.04

---

* Work supported by U. S. Atomic Energy Commission.

radiation lengths, internal radiative effects will predominate over radiation straggling in the target. (3) Single lens spectrometers of high performance become very massive. For these reasons the spectrometers

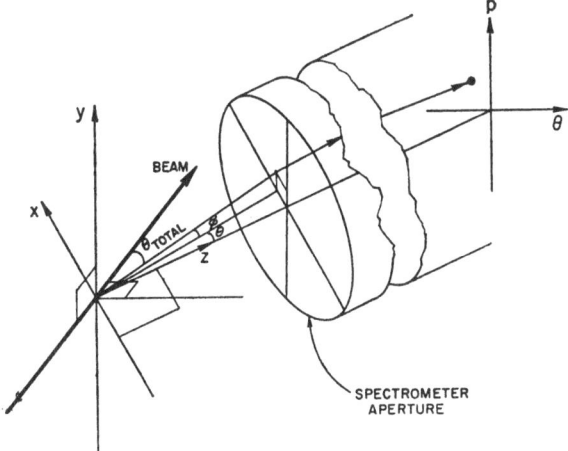

Fig. 1. "Pure" spectrometer

described here represent a convergence of the principles of the momentum analysers of low energy nuclear physics and the beam transport systems of high energy physics.

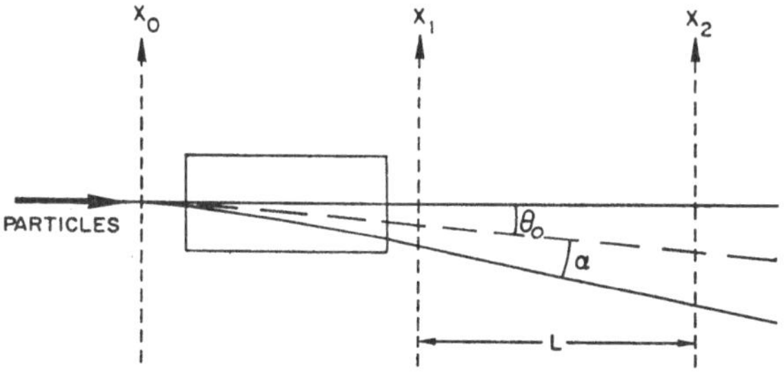

$$\delta_p = \frac{X_2 \sin\alpha - X_1[\sin\alpha + (L/\rho)\cos\alpha] + X_0(L/\rho)}{L(1-\cos\alpha)}$$

$$\theta = \frac{X_2 - X_1}{L} + \frac{X_1 - X_0}{\rho}\left(\frac{1-\cos\alpha}{\sin\alpha}\right)$$

Fig. 2. "Pure" hodoscope

What I might call a "pure" spectrometer will transform a given pair (θ, p) of the kinematic variables at the production point into a single resolution element (Fig. 1). On the other extreme we might consider a hodoscope arrangement determining a coordinate X_0 preceding a magnet bending the central particle orbit through an angle α at a radius of curvative ρ, followed by two hodoscopes separated by a drift distance L (Fig. 2). The linear transfer matrices give the expression for $\Delta\theta$ and

Δp shown on Fig. 2, assuming θ and X to be in a single plane; here Δp and $\Delta \theta$ are the deviations in momentum and production angle from the central setting. Calculations based on this type of geometry places extremely severe limits on the multiple scattering permitted in the hodoscopes if a resolution δp in the 0.1% range is required. Moreover, at least for poor duty cycle linear accelerates, the accidental rate would be prohibitive for most processes.

Naturally a whole continuum of arrangements ranging from a "pure" spectrometer to a pure "hodoscope" arrangement combined with a bending magnet of arbitrary focusing property are possible.

As example of the "pure" spectrometer we will here consider primarily "line-to-point" focusing spectrometers for use with energy electrons and photons which disperse θ and p at right angles; for mechanical reasons θ is dispersed horizontally and p vertically. In principle the same objective can also be met by a mixed system using point-to-point focusing both in the vertical and horizontal plane; the production angle θ can then be retrieved by an additional plane of hodoscopes following the focal plane. A coincidence between counter arrays at these two foci is thus required, with consequent higher accidental background rate when "structure" in the center-of-mass system is to be observed above a continuum. This alternative generally seems to offer little advantage over the "pure" spectrometer, but may be simples mechanically. In this class, is a version of the "Schrägfenster" spectrometer of Desy [1] in which $\Delta \theta$ is measured at an intermediate focus and a linear combination of $\Delta \theta$ and Δp is observed at a secondary focus.

An example of a "pure" spectrometer located in a horizontal plane is the rotated quadrupole spectrometer [2], described previously. In this instrument θ is first focused in a horizontal plane, but this focal position is imaged into a vertical coordinate through a "mixing" quadrupole. This system has been analyzed in detail to second order in all input coordinates but the complexity of alignment and of other adjustments made it less attractive than the other alternative systems which I will describe.

At high energies we are thus led to consider the family of multiple element spectrometers dispersing momentum vertically but providing line-to-point focus horizontally. At SLAC two such instruments are being designed; Table 1 gives a comparative table of parameters. In order to ascertain that the angular and momentum resolutions meet requirements, we have to be concerned with both first and second order optics. First order optics enters only to determine the momentum resolution δ_p: the dispersion must be sufficiently large so that the vertical magnification times the target height does not exceed the momentum resolution interval. This condition fixes in essence the required total size of the bending magnet, or more precisely the total available bending magnetic flux. In second order the situation is more complicated. If we consider the total initial variables $(X, \theta, y, \varphi, z, \delta_p)_0$ then following the formalism developed by K. L. Brown [3], we can write an aberration matrix for each magnetic unit of the spectrometer composed of "geometric" matrix elements such as $(X_{\text{final}} | X_0^2)$, $(X_{\text{final}} | X_0 \theta_0)$ etc. and including also "chromatic" terms $(X_{final}|\theta_0 \delta_p)$, $(X_{final}|X_0 \delta_p)$ and the like. We can then combine the resultant second order transfer matrices from each

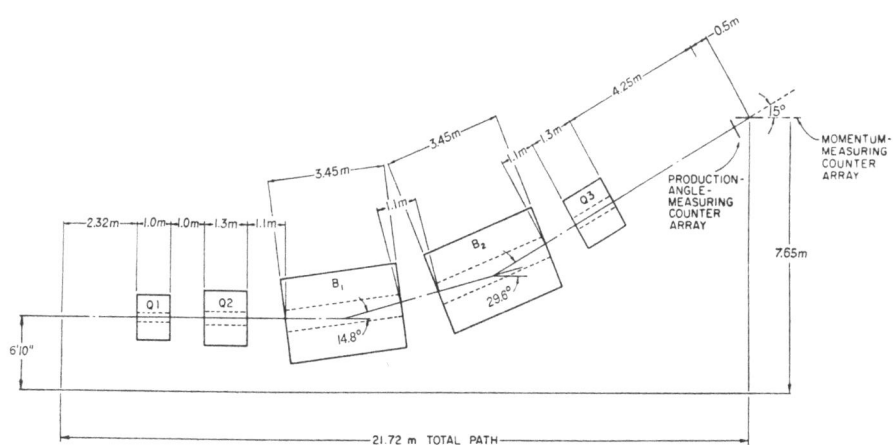

Fig. 3. Magnetic components of 8 GeV/c spectrometer

magnetic component and thus arrive at a program for obtaining both first and second order optics for the complete compound system. The

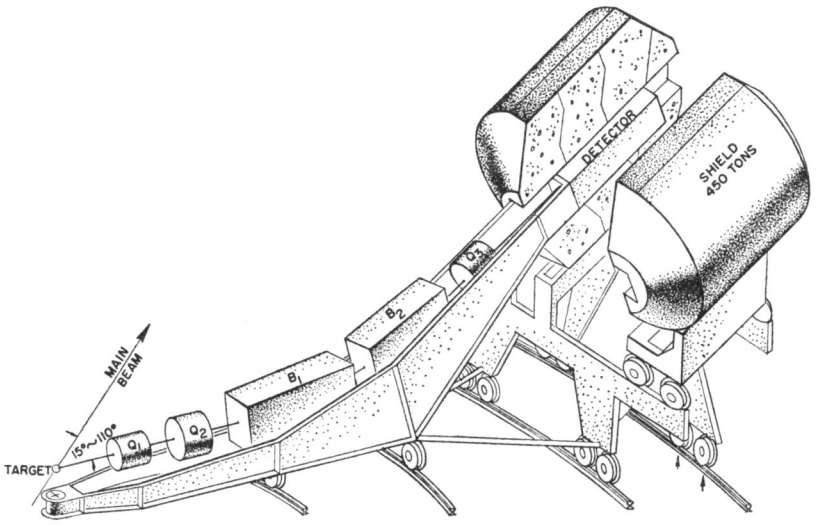

Fig. 4. Pictorial view of 8 GeV/c spectrometer

spectrometers in question were designed by using such a program, using essentially a trial and error procedure, guided by certain insight into the second-order effects. The resultant system for the 8 GeV instrument was configured as shown in Fig. 3. A physical picture of the instrument is shown in Fig. 4. Fig. 5 shows the "thin lens" equivalent of the instrument, showing how the vertical point-to-point and horizontal line-to-point focusing is achieved. The resultant actual orbits are shown in Fig. 6. The system was optimized for maximum solid angle meeting the resolution requirements of Table 1. Table 2 shows the overall transfer

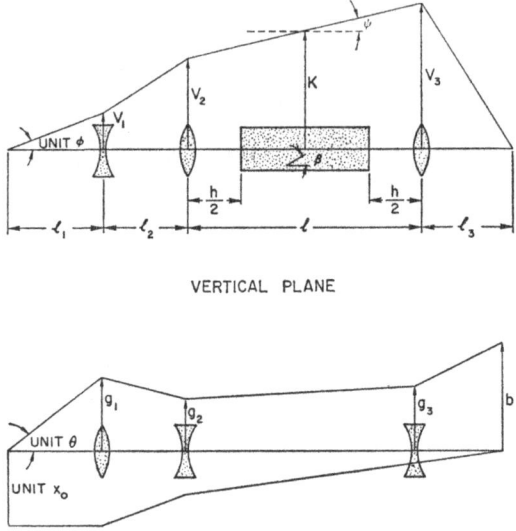

VERTICAL PLANE

HORIZONTAL PLANE

Fig. 5. Thin lens equivalent of 8 GeV/c spectrometer

matrix (angles in milliradians; distances in cm). We note that the desirable dispersion characteristics listed in Table 1 are approximately met. To minimize 2nd order coefficients, certain general properties are helpful: (1) to reduce chromatic aberrations in the vertical plane, the focusing orbits originating from a line target should have minimum slope during the bending magnet. This is indeed satisfied here, since the angle between the "sine-like" focusing orbit and the symmetry axis as shown in Fig. 6 can actually be chosen to be near zero. (2) Aberration terms like $(y | \varphi_0 \delta_p)$ correspond actually to a tilt in the focal plane through an angle ψ_y given by $- \tan \psi_y = (y | \delta_p) (\varphi | \varphi_0)/(y | \varphi_0 \delta_p)$. If we tilt the focal plane about a horizontal axis so that its plane is at 15° to the central ray,

Table 1. *Specifications of the SLAC 8 GeV/c and 20 GeV/c spectrometers*

	20 GeV/c	8 GeV/c
Momentum (max)	20 GeV/c	8 GeV/c
Momentum resolution	$\pm 0.05\%$	$\pm 0.05\%$
Solid angle acceptance	10^{-4} ster.	10^{-3} ster.
Momentum acceptance	$\pm 2\%$	$\pm 2\%$
Angular resolution	$0.3 \cdot 10^{-3}$ rad	$0.3 \cdot 10^{-3}$ rad
Angular range (production angle) . .	$\pm 4.5 \cdot 10^{-3}$ rad	$\pm 8 \cdot 10^{-3}$ rad
Azimuthal angle	$\pm 8 \cdot 10^{-3}$ rad	$\pm 30 \cdot 10^{-3}$ rad
Maximum target length (projected) . .	3 cm	10 cm
Minimum angle at which beam will miss instrument	$\approx 3°$	$\approx 12°$

then the overall effect on the final coordinates is given by Table 3. Since from the first order optics (Table 2) a 0.1% momentum excursion gives a vertical spread of 0.3 cm, and a 0.3 mrad angular spread a horizontal

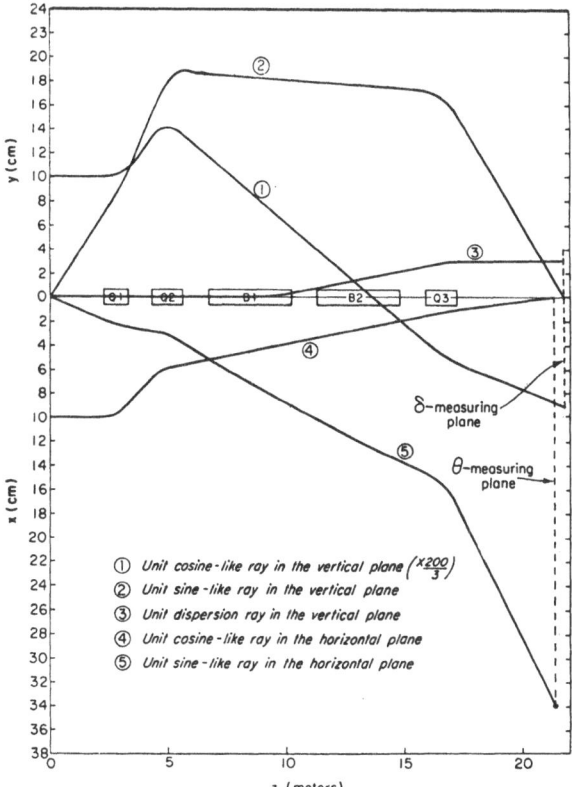

Fig. 6. Plots of computed orbits starting either with unit slope on the axis (sine-like orbit) or with zero slope at unit displacement (cosine-like orbit). The momentum dispersion of a ray starting along the central axis is also shown

excursion of 1.34 cm, we see that the resolution requirements are met. The result of these calculations has been verified at DESY by direct orbit tracing.

Table 2a. *Transfer matrix for vertical orbits of 8 GeV/c instrument*
(Units: Length in cm, angle in mr, momentum in %)

	x_0	θ_0	y_0	Φ_0	z_0	δ_0
x	—0.12	4.477	0	0	0	0
θ	—0.236	4.741	0	0	0	0
y	0	0	—0.947	0	0	—2.955
Φ	0	0	—0.845	—1.056	0	0.041
z	0	0	0.253	0.312	1	—0.382
δ	0	0	0	0	0	1

Table 2b. *Transfer matrix for horizontal orbit of 8 GeV/c instrument*
(Units: Length in cm, angle in mr, momentum in %)

	x_0	θ_0	y_0	Φ_0	z_0	δ_0
x	—0.000	4.240	0.000	—0.000	0.000	0.000
θ	—0.236	4.741	0.000	—0.000	0.000	0.000
y	—0.000	0.000	—0.905	0.053	0.000	—2.957
Φ	—0.000	—0.000	—0.845	—1.056	0.000	0.041
z	0.000	0.000	0.253	0.312	1.000	—0.382
δ	0.000	0.000	0.000	0.000	0.000	1.000

Table 3. *Equation giving the vertical and horizontal output coordinates in terms of all linear and all bi-linear combinations of the input variables of the 8 GeV/c spectrometer*

$$x = [4.24\,\theta_0 + 0.0108\,\theta_0\,\delta_0]$$

$+0.0405\,x_0\,\delta_0$	$(\pm 0.81 \text{ cm})$
$-0.000688\,\theta_0\,\Phi_0$	$(\pm 0.165 \text{ cm})$
$+0.000177\,x_0\,\Phi_0$	$(\pm 0.0531 \text{ cm})$
$+0.000947\,x_0\,y_0$	$(\pm 0.00142 \text{ cm})$
$-0.00053\,\theta_0\,y_0$	$(\pm 0.000635 \text{ cm})$

$$y = [-2.955\,\delta_0 + 0.00417\,\delta_0^2 + 0.000312\,\theta_0^2)]$$

$-0.947\,y_0$	$(\pm 0.14 \text{ cm})$
$+0.00126\,\Phi_0\,\delta_0$	$(\pm 0.076 \text{ cm})$
$-0.00368\,y_0\,\Phi_0$	$(\pm 0.017 \text{ cm})$
$-0.000154\,x_0\,\theta_0$	$(\pm 0.012 \text{ cm})$
$+0.0000197\,x_0^2$	(0.0020 cm)
$+0.0033\,y_0\,\delta_0$	$(\pm 0.0017 \text{ cm})$
$+1.08\cdot 10^{-6}\,\Phi_0^2$	(0.00097 cm)
$-0.00253\,y_0^2$	(-0.00006 cm)

The solution for the 8 GeV spectrometer was developed at the expense of the dispersed beam slanting upward. As the energy becomes larger this results in progressively more serious technical problems. For this reason the dispersion for the 20 GeV instrument was achieved by two bends of opposite sign, but with the vertical focusing orbits designed to make the dispersion of each magnet additive. The dispersed beam thus emerges horizontally.

Fig. 7 shows the lay-out of components and Fig. 8 the geometrical configuration of the 20 GeV instrument. The "double bend", required to produce a horizontal emergent beam was achieved at some sacrifice, but also produces some advantages. It can be shown [4] that the tangent of the angle between the final focal plane and the central ray of a high energy instrument using quadrupoles and uniform field bending magnets only is proportional to $(N + 1)^{-1}$ where N is the number of cross-overs in the dispersion plane. Hence the focal plane "tilt" angle for this instrument would be too small unless corrected by appropriate sextupole elements. Fig. 9 shows the "thin lens" equivalent of the "S-bend" instrument and Fig. 10 shows the resultant orbits as actually computed. Table 4 gives the first order transfer matrices of the intrument and Table 5 gives the final X and y coordinates in the focal plane as a function of both 1st and 2nd order terms. It is seen that all the specified resolutions appear to be met for the full aperture. An advantage of the double-bend instrument is that we have an intermediate focus in the dispersion plane as can be seen in Fig. 10. Hence we can place an intermediate slit near the symmetry plane of the instrument, thereby decreasing the flux of pole-scattered particles reaching the detector.

Several detectors are appropriate to instruments of this kind, depending on the experiment to be performed. Optimum utilization of data and minimum rates of accidental coincidences would indicate a two dimensional array (Fly's eye) set of counters. Although such an arrangement is

Table 4a. *Transfer matrix for vertical orbits of 20 GeV/c instrument*
(Units: Length in cm, angle in mr, momentum in %)

	x_0	θ_0	y_0	φ_0	z_0	δ_0
x	0.033	1.531	0.000	0.000	0.000	0.000
θ	—0.652	—0.066	0.000	0.000	0.000	0.000
y	0.000	0.000	0.927	—0.000	0.000	2.826
Φ	0.000	0.000	3.410	1.078	0.000	5.061
z	0.000	0.000	0.495	0.305	1.000	0.291
δ	0.000	0.000	0.000	0.000	0.000	1.000

Table 4b. *Transfer matrix for horizontal orbits of 20 GeV/c instrument*
(Units: Length in cm, angle in mr, momentum in %)

	x_0	θ_0	y_0	φ_0	z_0	δ_0
x	0.000	1.534	0.000	0.000	0.000	0.000
θ	—0.652	—0.066	0.000	0.000	0.000	0.000
y	0.000	0.000	0.757	—0.054	0.000	2.573
φ	0.000	0.000	3.410	1.078	0.000	5.061
z	0.000	0.000	0.495	0.305	1.000	0.291
δ	0.000	0.000	0.000	0.000	0.000	1.000

Table 5. *Equation giving output coordinates correct to second order in the input variables for the 20 GeV/c spectrometer*

$$x = [1.534\,\theta_0 + 0.0177\,\theta_0\,\delta_0]$$

$-0.0139\,x_0\,\varphi_0$	$(\pm 0.334 \text{ cm})$
$+0.0048\,\theta_0\,\varphi_0$	$(\pm 0.173 \text{ cm})$
$-0.105\,\theta_0\,y_0$	$(\pm 0.0708 \text{ cm})$
$-0.0878\,x_0\,y_0$	$(\pm 0.0395 \text{ cm})$
$-0.00254\,x_0\,\delta_0$	$(\pm 0.0153 \text{ cm})$

$$y = [2.826\,\delta_0 - 0.0421\,\delta_0^2 - 0.0167\,\theta_0^2]$$

$+0.927\,y_0$	$(\pm 0.139 \text{ cm})$
$-0.000836\,\varphi_0^2$	(-0.0535 cm)
$+0.0407\,y_0\,\varphi_0$	$(\pm 0.0488 \text{ cm})$
$-0.0029\,x_0\,\theta_0$	$(\pm 0.0391 \text{ cm})$
$+0.0042\,x_0^2$	$(\ \ 0.0378 \text{ cm})$
$+0.0291\,y_0\,\delta_0$	$(\pm 0.00873 \text{ cm})$
$+0.0653\,y_0^2$	$(\ \ 0.00147 \text{ cm})$
$+8.99 \cdot 10^{-5}\,\varphi_0\,\delta_0$	$(\pm 0.00144 \text{ cm})$

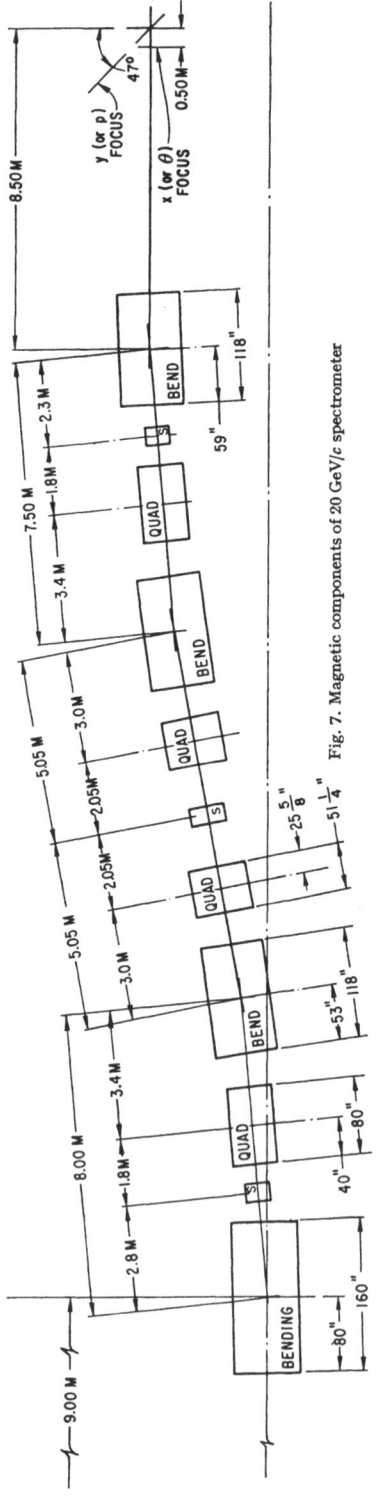

Fig. 7. Magnetic components of 20 GeV/c spectrometer

not excluded, the number of required elements (about 2000 for the 8 GeV instrument) would involve both very high costs and excessive mechanical complications if scintillators are used. A compromise is to use two grids of scintillators at right angles in a co-incidence matrix. Accidental coincidences at any one (θ, p) "bin" are than always accompanied by a coincidence elsewhere, and can thus be rejected by computer logic, unless the rates are too high. Another principle would be to use only a one-dimensional set of counters oriented in the (θ, p) plane to correspond to a given excitation energy in the center-of-mass frame. This arrangement is particularily suitable for small angle inelastic electron scattering and small angle photo-production studies. In all these cases the basic detection grid in the focal plane must be followed by a particle identification system; I will not discuss alternate designs of such system here. Suffice it to say that at large angles to the beam particle discrimination will pose severe problems if electrons are to be observed, since the expected pion to electron rates are extremely high.

Since spectrometers have a selective function — i. e., since they reject all but a small fraction of the phenomena occurring in a target — data reduction essentially concurrent to the experiment is needed to monitor the experiments. This need, in combination with the multiplicity of data channels and of the magnetic elements involved makes the use of an on-line computer highly desirable.

My second topic deals with the experimental methods of photon selection. Considerable progress in this field has been made since the last conference and I would like to mention some points of particular interest.

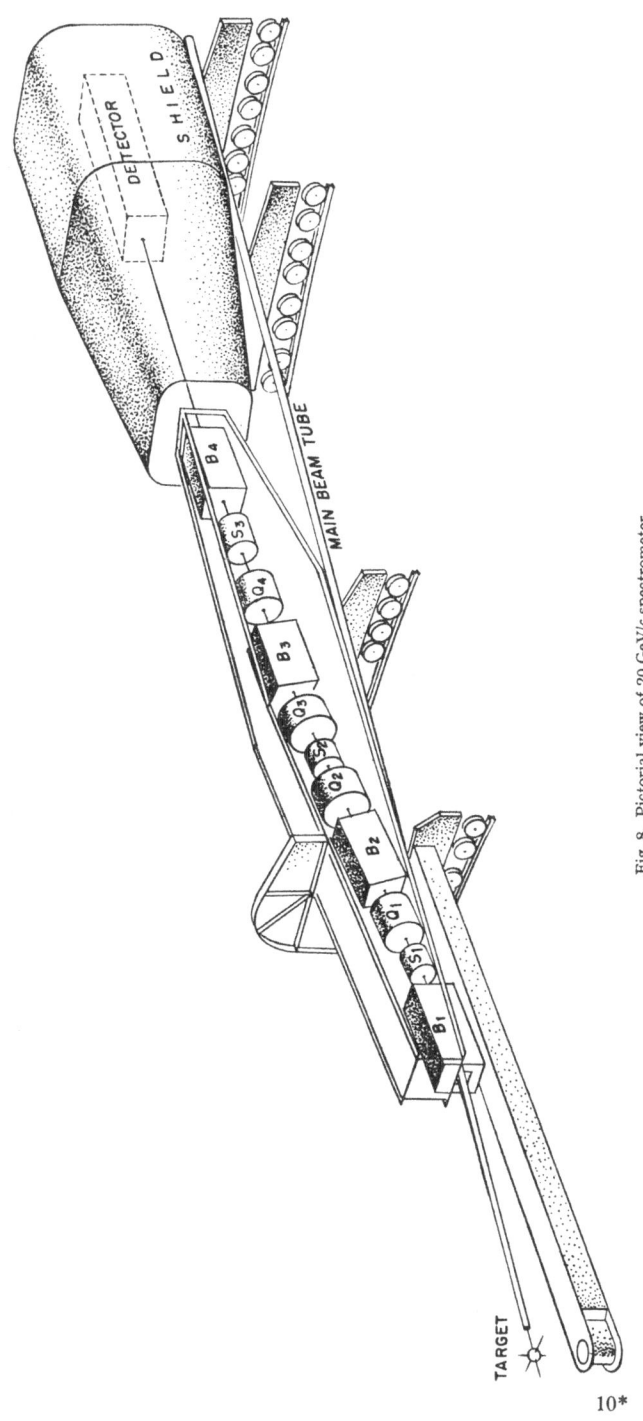

Fig. 8. Pictorial view of 20 GeV/c spectrometer

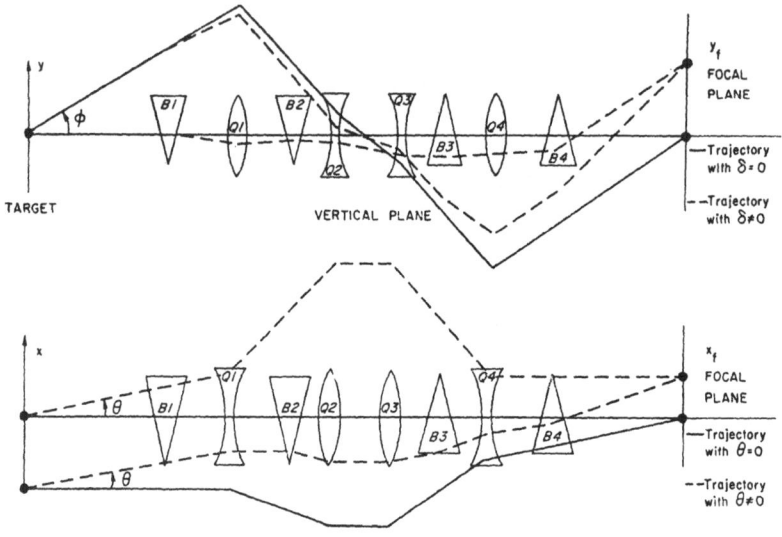

Fig. 9. Thin lens equivalent of 20 GeV/c instrument

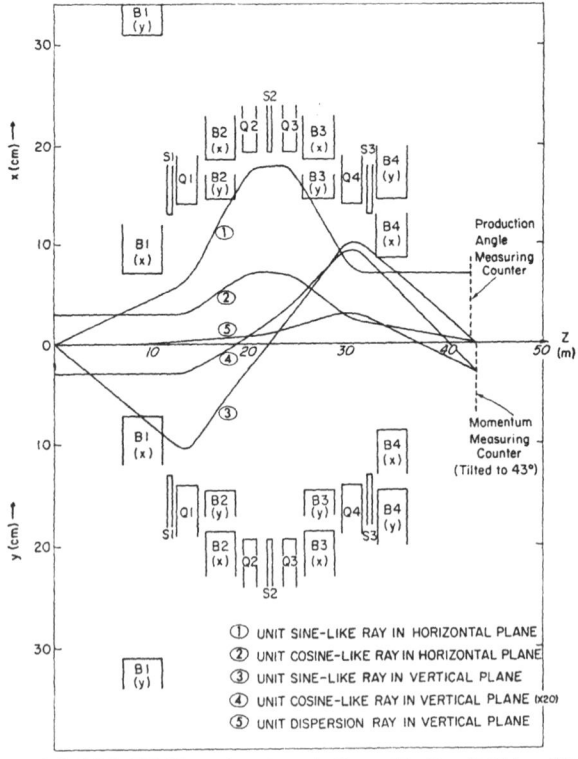

Fig. 10. Plot of computed orbits in 20 GeV/c spectrometer under the combination of initial conditions used in Fig. 6

High-energy-photo-reactions are usually produced in a bremsstrahlung beam filtered through a low-Z absorber to reduce the low-energy photon component. Four methods have been proposed or are under study to further "monochromatize" the photon beam. These are:

(a) The use of "tagged photons" — i. e., observing the photo-induced process in coincidence with a recoil electron from the radiator.

(b) The use of photons produced by Compton scattering of photons in a laser beam from a high-energy electron beam.

(c) The use of photons radiated from a monocrystal bombarded by high-energy electrons.

(d) The use of photons from annihilation of energetic positrons in flight.

The use of "tagged" photons is an old idea but is limited in application due to the problem of the low data rates required to avoid accidental coincidences.

The idea is originally due to KOCH at Illinois and CAMAC at Cornell. It was used extensively first by McDANIEL and collaborators [5].

PINE and others have recently "tagged" photos at the California Institute of Technology to obtain calibration sources for counters and other low counting rate applications. DRICKEY at SLAC has shown that even at a poor duty cycle a beam useful for spark chamber events can be obtained. E. g., for a typical arrangement an event rate of 2500 per hour can be obtained for a 100 μbarn cross section at an accidental coincidence background of 25%. Experiments using "tagged" photons for total cross section measurements of photons are in progress at CEA [6].

The method of Compton scattering of laser photons has been proposed both in ARMENIA [7] and in the U. S. [8]. Both groups calculated possible intensities, energy spectra and polarizations which might be obtained. Table 6 indicates typical values computed by MILBURN. There is little hope that this method can lead to photon beams of monochromaticity sufficient such that the primary energy would be a useful kinematic constraint in the analysis of a photon induced event; at best it can be hoped that the background of unwanted photons be reduced. A further problem in the application of the method is that the available photon energy is only a fairly small fraction of the electron energy, unless intensive laser sources in the very short optical wavelengths can be developed.

Table 6. *Table of energy and polarization of ruby laser photons back-scattered by electrons of initial energy E*

Electron Energy E	Peak Output Photon Energy	Peak Polarization
1.02 GeV	28 MeV	1.00
2.92	216	1.00
4.16	426	0.99
4.60	515	0.99
5.11	628	0.99
5.48	715	0.99
5.84	806	0.99
6.21	903	0.99
6.57	1.00 GeV	0.99
8.76	1.69	0.98
11.69	2.83	0.96
20.80	7.55	0.91
41.60	22.10	0.77
58.40	35.90	0.67

Recently the Tufts University — CEA group [9] has observed the expected process successfully. Photons of 0.85 GeV were produced by scattering of the 6943 Å quanta from a ruby laser on the 6 GeV electron beam. The laser beam is reflected by a mirror to form a light beam in the direction opposite to the circulating electron beam of CEA. The back-scattered photons were observed in a Cerenkov Counter. The experiment at this time is mainly an "existence proof" rather than the source of a useful beam. The experiment has shown that the intensity of the back-scattered light was proportional to the product of the electron beam and laser beam intensities. The actual number of photons was probably 15—20 for each 0.2 joule of laser incident pulse.

The scattering of laser light by fast electrons has also been observed in the USSR [10]; the laser photons were scattered on a 600 MeV beam, resuling in 7 MeV back-scattered photons. The principal motive of experiments of this kind is not as much their intrinsic interest, since these processes make very low four-momentum transfer, but their potential application to beam formation. Milburn has studied practical installations, as have the Armenian workers. Milburn calculates a photon output for a feasible geometry at SLAC of $3 \cdot 10^4$ photons per laser joule per 1 μsec pulse of the accelerator; the number could be increased by an order of magnitude for 0.1 μsec pulses if the pulse timing precision can be achieved. Thus the laser beam is equivalent to about 10^{-8} radiation length of radiator in the electron beam; residual gas background can therefore become appreciable, but can be controlled. Useful intensities are thus possible, but the question of achieving useful monochromatization is quite difficult. The characteristic angle ($\gamma_{\text{effective}}^{-1}$) defining the angular variation of photon energy k with angle θ in the equation $(k/k_{\max}) = 1/[1 + (\gamma_{\text{eff}}\theta)^2]$ is given by $\gamma_{\text{eff}} = \gamma/[1 + 4(\gamma k_i/m c^2)]^{1/2}$ where k_i is the laser photon energy; for 20 BeV electrons $\gamma_{\text{eff}}^{-1} = 3 \cdot 10^{-5}$ radians; monochromatization to 1% thus requires control of the angle between the electron beam and the outgoing photon to $3 \cdot 10^{-6}$ radians. Even at SLAC this is a formidable problem; it is likely that monochromatization for reducing background can be achieved, but that a line width sufficiently narrow to produce a constraint in the analysis of track chamber events is exceedingly difficult. Thus possibly the most promising application of laser beam photons is based on their high polarization rather than their spectrum. However, much effort is clearly required to reduce this technique to practice.

It has been pointed out by Mozley and de Wire [12] that the intensity peaks observed in the bremsstrahlung spectrum produced by electrons on monocrystals could be used as a possible candidate for a monochromatic source. The beautiful experiments by Diambrini and co-workers [13] demonstrated that the spectrum radiated by 1 GeV electron on a single diamond crystal had the theoretically predicted "sawtooth" shape. Mozley and de Wire pointed out that the "sawtooth" shape could be converted into a series of peaks by restricting the angular tolerance between the incident electron beam and the outgoing photon beam to angles of order γ^{-1}. We are here thus again forced to angles in the 10^{-6} radian range and thus to very large distances and constraints on incident electron phase space. Hence the problems of practical realization are not dissimilar to those involving laser beams. However,

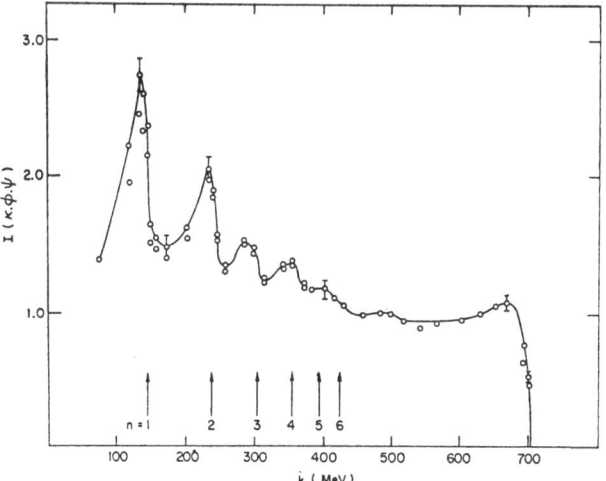

Fig. 11. Plot of bremsstrahlung intensity by 720 MeV electrons incident on Silicon as observed by KATO et al. [14]

a radiation length of 10^{-4} can be used without excessive multiple scattering and thus higher intensities are feasible than with laser beams. The enhancement observed by DIAMBRINI et al., should increase rapidly with energy, provided we make the dubious assumption that the calculations based on Born approximation remain valid. In diamond the beam heating effects may force operation at less than maximum incident electron beam. However, recently a group of Japanese workers [14] have obtained energy peaks from single Silicon crystals similar to those obtained earlier in diamond by DIAMBRINI and co-workers (Fig. 11). The measurements were made with incident electrons of 720 MeV; therefore the enhancement is less than that observed by the Italian group. Nevertheless, KATO et al. were able to demonstrate the sharpening of the intensity dependence on the angle between the incident beam and the crystal axes as the collimation of the outgoing photons was tightened. As the result of the successful observation of the spectral fine structure of high energy bremsstrahlung from Silicon, it is probably worth while to re-open the question of using crystaline radiators in practical application. Nevertheless, the angular precision required and the incoherent background make this method not too attractive.

The radiation from monocrystals is highly polarized; such polarized bremsstrahlung has been used for photoproduction studies [15]. This method gives results in agreement with the earlier work on polarized beams from noncrystalline targets, but permits thicker radiators.

The method for photon beam monochromatization which looks most attractive for hydrogen bubble chambers is the use of annihilation of positrons in flight. The photon energy k observed at an angle θ from positrons of energy $\gamma\mu$ is given by $k/\mu = (1 + \gamma)/(1 + \gamma^{-1})\,(1 + \theta^2\gamma^2)$.

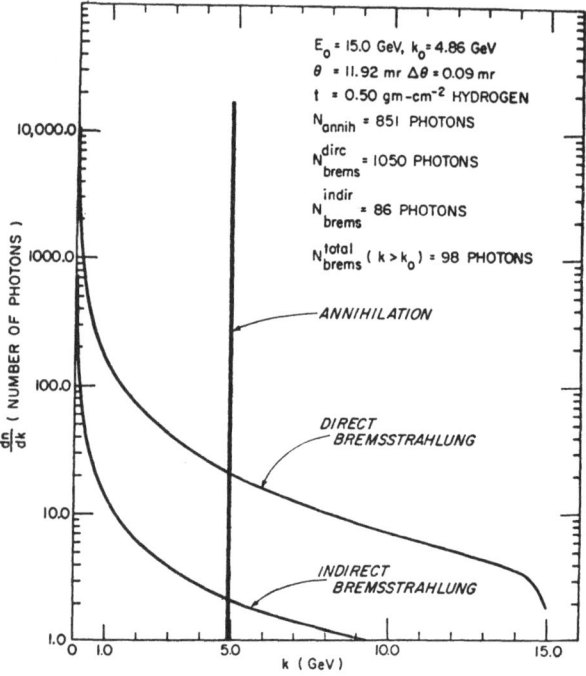

Fig. 12. Plot of the total γ ray spectrum produced by 15 GeV positrons in hydrogen observed at an angle of 11.92 milliradian

The angular distribution of the cross section per unit solid angle of the emitted photons varies as θ^{-2} for $\theta\gamma \gg 1$, while the cross section for the competing bremsstrahlung process varies as θ^{-4}. Hence the ratio of annihilation to bremsstrahlung can be reduced by increasing θ until the photon energy k decreases substantially below the incident energy; this occurs if $\theta \sim \gamma^{-1/2}$ according to the above equation.

The use of monochromatic photon beams by positron annihilation has been used extensively in low energy photo-nuclear reactions. Its use for high energy electrons was first discussed by Cassels [16]. Further, more detailed, calculations are due to Guiragossián [17]. The problem here is to choose parameters, in particular the photon angle, such that the ratio of photons in the peak to that in the continuum is acceptable without degenerating the energy excessively. Fig. 12 shows a typical curve pertaining to such a compromise and the parameters pertaining to it. The accuracy of monochromatization is given by $dk/k = (k/\mu)\,\theta\,d\theta$. Hence the photon direction must be accurate to $\sim 10^{-4}$ rad for a 1%

photon accuracy. Hence from this point of view this method is superior to the ones previously discussed. With the $\sim 1\ \mu A$ of positrons expected at SLAC, intensities are amply sufficient for bubble chamber use.

There exists an interesting alternative in combining the beam tagging and monochromatization techniques [18]. The two γ-rays from positron annihilation will have a fixed geometrical relationship to one another. Therefore, if one γ-ray enters the bubble chamber, the other one can be located in a conventional spark chamber with lead plates. Spatial resolution should be adequate to determine which γ induced event in the bubble chamber corresponds to a matching photon in the spark chamber.

All these methods of monochromatizing real photons are, of course, in competition with the analysis of virtual photon processes induced by inelastic muon or electron scattering. I will not make such comparisons here; several authors, including L. HAND, D. DRICKEY, M. PERL, J. TINLOT and several others have analyzed possible future virtual photon experiments using track chamber techniques. At the risk of over-simplifying the description of a complex series of alternatives, I might summarize that the real photon techniques described above appear superior particularly if background problems are considered. However the analysis of inelastic events gives of course additional information such as on polarization dependence, on the contribution of longitudinal matrix elements and on the pertinent form factors.

References

[1] KERN, W., and K. G. STEFFEN: DESY/E 3, 4, 5 (1962); HUBNER, M. L.: Private communication

[2] PANOFSKY, W. K. H., D. COWARD, and K. L. BROWN: A High-Resolution 20 BeV Spectrometer Resolving Momentum and Production Angle Independently, Proceedings of the International Conference on High-Energy Physics, Dubna 1964

[3] BROWN, K. L.: First- and second-order aberration coefficients of wedge magnets. The general first- and second-order theory of beam transport optics, Part II: Reduction of the general first- and second-order theory to the case of the ideal wedge magnet, SLAC Internal Report, Stanford Linear Accelerator Center, Stanford University, Stanford, California (1963—1964)

[4] BROWN, K. L.: Private communication

[5] WEIL, J. W., and B. D. McDANIEL: Phys. Rev. 92, 391 (1953)

[6] WILSON, R.: Private communication

[7] ARUTYUNIAN, F. R., and V. A. TUMANIAN: Physics Letters 4, 176 (1963); Zh. Eksp. Theor. Fiz. 45, 312 (1963) [JETP 18, 218 (1964)]

[8] MILBURN, R. H.: Phys. Rev. Letters 10, 75 (1963)

[9] BEMPARAD, H., and R. H. MILBURN: Tanaka CEAL-1014; January 1965

[10] KULIKOV, O. F., Y. Y. TELNOV, E. I. FILIPPOV, and M. N. YAKIMENTO: Phys. Letters 13, 344 (1964)

[11] MILBURN, R. H.: SLAC Report, August 1964; in preparation

[12] MOZLEY, R. F., and J. DE WIRE: Nuovo cimento 27, 1281 (1963)

[13] BARBIELLINI, G., G. BOLOGNA, G. DIAMBRINI, and G. P. MURTAS: Phys. Rev. Letters 8, 112 (1962) and 8, 454 (1962)

[14] KATO, S., T. KIFUNE, Y. KIMMURA, M. KOBAYASKI, K. KONDO, T. NISHIKAWA, H. SASAKI, K. TAKAMATSU, S. KIKUTA, and K. KOHRA: J. Phys. Soc. Japan 20, 303 (1965)

[15] BARBIELLINI, G., G. BOLOGNA, J. DE WIRE, G. DIAMBRINI, G. P. MURTAS, and G. SELT: Supplements Nuovo cimento II 4, 666 (1964)

[16] CASSELS, J.: Some aspects of target area design for the proposed Stanford two-mile linear electron accelerator, SLAC. M. Report 1960

[17] GUIRAGOSSIAN, Z. G. T.: Almost monochromatic photon beams for a hydrogen bubble chamber, 1963, SLAC Internal Report, Stanford Linear Accelerator Center, Stanford University, Stanford, California, and BALLAM, J., and Z. G. T. GUIRAGOSSIAN: Proceedings of the International Conference on High-Energy Physics, Dubna, August 5—15, 1964

[18] DRICKEY, D.: Tagged Photons, SLAC Internal Report, Stanford Linear Accelerator Center, Stanford University, Stanford, California (1965)

Prof. Dr. W. K. H. PANOFSKY

Stanford Linear Accelerator Center
Stanford/California

The Use of Bubble Chambers and Spark Chambers at Electron Accelerators*

Karl Strauch

Lyman Laboratory of Physics
Harvard University,
Cambridge, Mass.

A. Introduction

When I was asked a few months ago to give a talk on the problems of the use of bubble and spark chambers at electron accelerators, I accepted with great pleasure. As I sat down to prepare my remarks, I began to feel more and more like those experts concerned with the great and apparently difficult problem of the "Affluent Society" – what to do with all the leisure time which is becoming available thanks to the increased use of advanced machines and computers. Of course for the physicist the problem does not exist – his emphasis has simply shifted from taking care of girls to debugging computers, automatic scanning and measuring equipment. But I feel quite confident that even for the general public the problem will take care of itself with or without too much learned discussion of the philosophy of the problem.

Similarly before the completion of multi-GeV electron accelerators, there was much discussion about the usefulness of bubble chambers at these new facilities. Spark chambers did not enter into these discussions, they had not been invented. It has been demonstrated by the work at CEA and recently at DESY that hydrogen bubble chambers work very well with photon beams, and that important results are obtained which are difficult to obtain by any other technique. So the main problem has taken care of itself.

The first part of my talk will be concerned with those aspects of the use of bubble chambers which differ at electron and proton accelerators. In practice this means the use of photon and electron beams, since secondary beams of strongly interacting particles do not know in what

---

* Work supported by U. S. Atomic Energy Commission.

type of accelerator they have been created. Some of the same problems are encountered with spark chambers, and these will be discussed in the second part of this talk. Finally the use at accelerators of the recently developed spark track chambers (or wide gap chambers) will be described.

The ideas, techniques and results that I shall discuss were developed by many individuals — alone, together or independently. Under these circumstances it is difficult to always give proper credit. I apologize in advance for any oversights which I am bound to make.

B. The present use of bubble chambers at Multi-GeV electron accelerators

Both at CEA and at DESY [1] a photon beam has been passed through 30 cm and 80 cm hydrogen bubble chambers respectively. To insure the visual observation of the interaction vertex, the beam intensity was kept low to a $Q \cong 50$ equivalent photons and low energy photons ($E_\gamma \lesssim 20$ MeV) were removed by a beam hardener. A double conversion method was used to produce a beam of sufficiently low intensity. This beam does not necessarily represent an optimum design, but one that worked. The Cambridge and German collaborations have obtained one nuclear event per 130 and 30 pictures respectively.

Two features of this work are worth emphasizing.

(1) The intensity and energy distribution of the incident photon beam is directly obtained from the number and energy of electron-positron pairs created in the chamber itself. The accuracy of cross sections are limited only by statistics and the knowledge of the pair production cross section in hydrogen.

(2) The bubble chamber is ideally suited for simultaneous operation of one or more additional experiments since at most a few beam pulses per second can be used.

Is there anything special in the analysis of the pictures and the extraction of the physics when photon beams are used? The *dominating feature* is the *continuous energy distribution* of the hardened bremsstrahlung beam. What are some of the consequences?

(1) As long as the beam is well collimated, the photon direction but *not* its energy are known. In reactions in which all final particles are observed in the chamber (i. e. $\gamma + p \to p + \pi^+ + \pi^-$ or $\gamma + p \to \Lambda^0 + K^+$) the kinematic problem is still well over-determined and the lack of knowledge of the photon energy results at most in a small reduction of the experimental accuracy.

However, if one neutral particle is emitted (i. e. $\gamma + p \to p + \pi^+ + \pi^- + \pi^0$ or $\gamma + p \to \Lambda^0 + K^+ + \pi^0$) the lack of knowledge of the photon energy has a much more serious effect. It becomes impossible to calculate the mass of the neutral particle i. e. to identify it. This is a serious handicap since it makes it impossible to separate and identify reactions in which one neutral particle only is emitted from those in which more than one neutral particle are produced. All the kinematic analysis can say is that at least one unknown neutral particle is emitted in addition to those that are seen.

The usual way of handling such an event is to assume the neutral particle to be a π^0 and to calculate the kinematics accordingly. Cross section-measurements [1] suggest that this is a good assumption below about 2.5 GeV, but becomes progressively poorer at higher energies. Thus a three prong event which does not check out double pion production is assumed to be $\gamma + p \rightarrow p + \pi^+ + \pi^- + \pi^0$ or $n + \pi^+ + \pi^- + \pi^+$ although more than 1 π^0 might well have been emitted.

This inability to separate single from multiple neutral particle emission not only makes it difficult to separate certain reaction channels, it also means that the photon energy calculated for multiple neutral particle events is too low. Any resonant state decaying with the emission of one unseen particle (i. e. ω^0, Y*) becomes more difficult to identify since its mass peak will ride on a higher background of events that are not properly identified.

The lack of knowledge of the neutral mass represents the most serious problem in the present experiments — some of the other problems will be taken care of by hard work!

(2) A wide photon energy range is studied at one time. This is very useful to get an overall view of the various photoproduction processes, but it may cause a very energy-dependent effect not to be observed until a vast number of pictures have been processed. Of course with charged particle beams a similar effect might be missed until many beam energies have been covered.

(3) Because of the nature of the bremsstrahlung beam, the number of high energy photons is relatively small. Since in addition the cross section for visible events tends to decrease with increasing photon energy [1], several low energy events must be processed for each high energy event that is observed.

(4) Some of the common tools of high energy physics lose much of their usefulness. Because of the continuous energy distribution of the incoming photons, Dalitz plots become complicated non-uniform distributions which are hard to analyze.

C. Possible photon beam improvements

How is it possible to increase the effectiveness of bubble chambers exposed to photon beams to produce physics? Up to now the main emphasis at CEA and DESY has been to make the chambers work and to match them to the accelerator. However, it seems worthwhile to consider the usefulness for bubble chambers of the several methods that have been proposed and studied to (a) enrich or concentrate the photon energy distribution into narrower and narrower regions; (b) create polarized photons; and (c) know the energy of the photon that causes a given interaction. The proposed methods usually combine more than one of these features in varying degrees, and these features are functions of the electron energy and the available geometry. The usefulness of these methods to spark chambers will also be noted as we go along. Only the last method ("tagging") will be considered in some detail since the others are discussed by Professor PANOFSKY in the previous paper.

1. Electron-Positron annihilations in flight. When a positron passes through matter, it can annihilate in flight with a target electron into two photons. For a given angle θ of the photon with respect to the incident positron, a unique photon energy value is obtained. As θ increases, the energy decreases and the number of photons decreases as $1/\theta$. By suitable collimation it is thus possible to obtain a monochromatic photon beam by this process. Two sources of background radiation occur: the ordinary bremsstrahlung which decreases in intensity as $1/\theta^3$ and an indirect bremsstrahlung in which the positron has a large angle scattering before it emits bremsstrahlung. The possibility of obtaining almost monochromatic photon beams by this method has been studied extensively by BINNIE [2] and by others at SLAC [3, 4]. They conclude that it is possible to produce a narrow energy line which rides on top of a bremsstrahlung continuum. The recent suggestion of DRICKEY (reported by Professor PANOFSKY in the previous paper) to identify a photon in the narrow line by observing its spatially correlated photon partner, promises to increase the usefulness of this method. In addition by measuring the energy of the recoil positron which emits a "background" bremsstrahlung photon it is possible to know the energy of the interacting photons that are not included in the line. This will be discussed more fully in the section on tagging. The low intensity of this type of beam makes it ideal for the bubble chamber, much less so for a spark chamber experiment.

2. Coherent bremsstrahlung in crystals. The coherent bremsstrahlung emitted by electrons passing through a crystal was first observed by DIAMBRINI and collaborators and has been studied extensively since by this group at FRASCATI [5]. The relative number of high energy photons is increased and they are polarized. While the enrichment into a single narrow line is less pronounced than for the positron annihilation method, the much larger intensity makes this method highly suitable for spark chamber work.

3. Compton scattering of laser beams. When an intense laser beam is directed on a beam of multi-GeV electrons, the doubly Doppler shifted scattered photons have energies that extend to a fraction of the electron energy. The photon energy spectrum tends to be flat peaking somewhat at its highest energy. The use of a polarized laser beam makes it possible to have a highly polarized high energy end of the photon spectrum. This method of photon beam enrichment was suggested independently by MILBURN [6] and ARUTYNUYAN and TUMANYAN [7, 8] and has been shown experimentally to be practical [9].

Professor MILBURN has made a careful examination of the practical aspects of the creation of such photon beams at CEA and at SLAC [10]. He will report his experimental results and conclusions in detail at this conference [11].

The Compton scattering method appears to be the most promising way of obtaining an enriched photon spectrum (flat instead of $1/E$) with nearly complete polarization at the high energy end.

4. Polarization of bremsstrahlung. By selecting a small portion of an ordinary bremsstrahlung beam, polarized photons are obtained. This method has been used by MOZLEY and collaborators [12, 13] with $15-20\%$ calculated polarization. This method of polarization provides no enrichment, but produces beams of sufficient intensity for spark chamber work.

5. Tagging of recoil electron in bremsstrahlung. This is the oldest method invented originally by KOCH at Illinois and CAMAC at Cornell to measure the energy of the photon from a bremsstrahlung beam which causes a given event. It was used extensively by McDANIEL and collaborators at Cornell. The primary electron beam of energy E_0 is passed through a thin converter ($\sim .01$ rad length) and the energy E_e of the electron that has radiated is measured. This electron is put into time coincidence with some product of the event under study, so that each event is tagged with the energy $E_0 - E_e$ of the photon responsible. This method was used at CEA by CALDWELL et al. [14], and extensive photo-production experiments by this method with spark chambers are in preparation both at CEA and DESY.

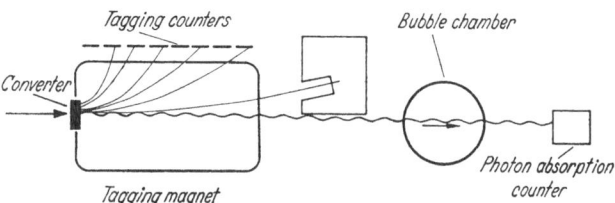

Fig. 1. Schematic drawing of proposed bubble chamber tagging installation

With the duty cycle available with circular machines accidental coincidence limits the use of tagged photons to about $5 \cdot 10^3$ equivalent photons/1 millisec pulse. This intensity is a factor 10^6 lower than is available with ordinary bremsstrahlung beams, and large solid angle detectors are required. This is why spark chamber set ups are well matched to tagged photon beams. It is more difficult to use this method with the poor duty cycle of SLAC.

Is it possible to match tagged photon beams to bubble chambers? We have recently considered several possible schemes in Cambridge and the situation looks encouraging. The simplest scheme is illustrated in Fig. 1. The tagged photon beam is passed through the bubble chamber, and then absorbed in a photon counter. For each bubble chamber expansion, a record is kept of the energy of the recoil electrons that are in anticoincidence with the photon absorption counter. Thus for each picture there will exist a list of energies of photons that have interacted between the converter and the absorption counter. These interactions are of three types: (a) electromagnetic pairs in the hydrogen. These pairs can be measured directly and the corresponding photon energies eliminated. (b) nuclear events in the hydrogen. Neglecting background (c) the energy value left after energies (a) have been removed will be the energy responsible for the event. (c) electromagnetic and nuclear events produced in or near the back window. It appears feasible to make this background relatively small so that usually no more than one possible energy value will be left for the nuclear event. Since the recoil electrons are not in coincidence with any reaction product, the background in the tagging counters must be reduced by the use of telescopes and/or energy sensitive counters.

Other schemes that have been considered are more elaborate, but do not rely on the operation of an anticoincidence photon absorption counter. One scheme uses a set of very fast sweeping magnets to sweep a tagged pencil photon beam across the window area. In this way a correlation is established between the position of the photon beam in the bubble chamber and the time at which a recoil electron of a certain energy was observed. A similar correlation is established in another scheme in which a large electron beam impinges on a converter which consists of a counter matrix. Coincidences between the converter counters and the tagging counters then establish the desired correlation. In either scheme, or any mixture thereof, a library tape is established in which each picture has a list of photon energies and corresponding positions. Reference to this tape will then establish the photon energy of a nuclear event in a given location. For beam intensities that have been used with bubble chambers so far and not-over-elaborate set ups, indications are that the last two schemes will result in a somewhat more ambiguous energy determination than the first one which uses the anticoincidence counter. The most promising scheme involves a combination of anti-coincidence counter with beam sweep with a pulsed magnet.

It thus appears feasible to use tagged photon beams in connection with bubble chambers. The combination of tagging and annihilation in flight looks particularly interesting. Passing a positron beam through a thin converter, it is possible to construct a photon beam containing a narrow energy line due to annihilation in flight and to identify these photons by their spatial correlation with the recoil photon observed in a spark chamber. This narrow energy line is superimposed on a bremsstrahlung spectrum. If the positrons are tagged, then the energy of the photons in the continuum are also known. Thus this combination produces a tagged photon beam enriched by a narrow photon line whose energy position can be changed by changing the angle of the photon beam with respect to the positron beam. The intensity of such a beam would match a bubble chamber, but appears very low for spark chambers.

6. Conclusions. Several methods exist to enrich the relative number of high energy photons (although at the expense of reducing the maximum photon energy), polarize photons or tag them with the correct energy. All of these methods are fairly elaborate and require development. After the present series of bubble chamber exposures with ordinary bremsstrahlung are terminated, the physics will decide which, if any, of the proposed schemes will be the most useful for a given specific study.

D. The use of spark chambers

Most uses of spark chambers at electron accelerators is not different from those at other accelerators. There is no point repeating here all of the well known characteristic features of multiplate spark chambers that make them so very useful. For some investigations such as

$$\gamma + p \rightarrow p + \rho^0$$
$$\rightarrow p + \omega^0$$

a knowledge of the photon energy is desirable or even necessary. As we have just seen, this can be accomplished best with sufficient intensity by the tagging method. Since this involves a considerable reduction in tagged photon intensity over normal beams, spark chamber detectors subtending large solid angles are indicated. These remarks apply to both the tagging of real photons which has been discussed, and to the tagging of virtual photons in electro-production in which the energy of the scattered electron is measured in coincidence with the event under observation. It might be noted that the short sensitive time and low repetition rate of hydrogen bubble chambers limit the possible tagged beam intensity to $10^{-3} - 10^{-4}$ times the value possible with spark chambers.

The recent development (principally in the U. S. S. R.) of wide-gap or "track spark chambers" is adding a new tool to those available to the experimentalist, which appears particularly promising where a large solid angle detection and accurate trajectory reproduction are desired. I have time here only briefly to review the operation of these chambers and to discuss those characteristics that are important in their operation at accelerators. A more complete discussion with bibliography can be found in the Proceedings of the Purdue Conference [15].

An ionizing particle passes through a noble gas such as Ne. It leaves a trail of free electrons along its trajectory. Application of a high electric field of the order of 10 kV/cm produces avalanches from which emerge streamers moving in the general direction of the electric field. It is only after streamers are formed that sufficient light is emitted for observation. If the electric field is now suddenly removed, the trajectory is seen as a series of luminous dots or lines, the streamers. In this *"streamer mode"* the light is limited and good depth of field photography is difficult [16, 17].

If the electric field is applied for a longer time and the trajectory makes an angle less than $\sim 45°$ with the electric field, streamers join to form a spark channel through which can pour a lot of energy producing a lot of light. This is the *"spark mode"*. I have been unable to find a "start" reference for this mode. It appears to have been a gradual development, perhaps originating in the work of CHARPAK at CERN. At any rate ALIKHANYAN et al. [18] are the first to report a curved spark track in a magnetic field. The main advantage of the streamer over the spark mode is that it has isotropic characteristics, but at the expense of reduced light intensity and more critical operating conditions. Fig. 2

Fig. 2. Cosmic ray tracks produced in 150 cm long × 35 cm deep × 40 cm wide spark chamber (multiple exposure)

shows cosmic ray tracks in a 150 cm long, 35 cm deep and 40 cm wide spark chamber built at Harvard for use at CEA. Fig. 3 shows two electromagnetic pairs produced in the walls of a 40 cm gap chamber

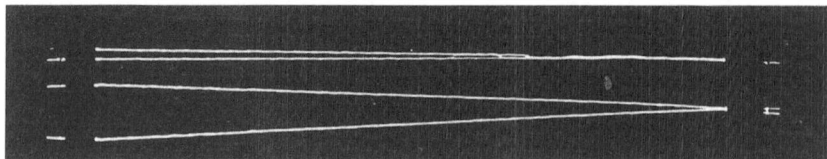

Fig. 3. Electromagnetic pairs produced in the wall of a 40 cm wide spark chamber by 6 BeV bremsstrahlung

exposed to a weak photon beam at CEA. Some very exciting work is being carried out at SLAC by BULOS, BOYARSKI, ODIAN and VILLA on making completely isotropic streamer chambers using very short pulses and an image intensifier to have sufficient light for photography.

The fundamental property which makes the track spark chamber so interesting is that the spark or streamer track reproduces the particle trajectory to a high degree of accuracy. Accurate curvature measurements are possible when the track spark chamber is placed in a magnetic field, with $\vec{B} \parallel \vec{E}$ for the streamer mode, and $\vec{B} \perp \vec{E}$ for the spark mode. This accuracy requires a *homogenous electric field* in both the streamer and the spark mode ($\vec{E}$ drift), and in addition a *homogenous magnetic* field in the spark mode because of the $\vec{E} \times \vec{B}$ motion.

This accurate reproducibility is due to the following:

(1) The spark or streamers are observed within gas, away from the electrodes where distortions occur.

(2) The displacement of track from trajectory depends on the detailed time history of the high voltage pulse and the condition of the gas. As long as all portions of the track see at all times the same electric field and are in the same gas, the displacement is coherent and no apparent curvature is introduced.

(3) The multiple scattering in 1 atm of neon is considerably less than in multiplate chambers of the same total length (radiation length of Ne at 1 atm is 30×10^3 cm).

(4) The track formation is so rapid that there are no effects of gas turbulence.

A second important property is the high efficiency for multiparticle events of the spark track chamber. This is due to the fact that long tracks involving many avalanches are observed. Fluctuations in the number and formation of avalanches are therefore small. However, if in the spark mode the chamber is tuned for observation of one or a few particles, a many-particle event will have dim tracks, a single-particle event will appear very bright. This effect should be much less important in the streamer mode, although it appears to exist there too.

What are the most important characteristics of these chambers for accelerator experiments? We will consider the following: accuracy of trajectory reproduction; sensitive time; stability of operation; maximum permitted background; ionization measurements.

Most of our own experience has been with the spark mode, and this is the one that will be stressed for that reason.

1. Accuracy of trajectory reproduction. By comparing the curvature of the spark tracks produced by one particle in two separate chambers located in a magnetic field, GARRON et al. [19] have found that the momentum of 1 BeV/c particle in 13 kg and a 40 cm wide chamber can be determined to $\pm 1.4\%$. (This number does not represent the limit of accuracy under these conditions, but one that was obtained.) To obtain this accuracy, care must be taken to prevent distortion of the spark track by non-uniform electric and magnetic fields. Similar accuracy should be obtainable in the streamer mode, but this remains to be shown in detail.

When placing a chamber near the poles of a magnet, the electric field is badly distorted. Unless special precautions are taken, no tracks or only very distorted ones will be observed. We have found the following shielding arrangement convenient and effective. Wires are placed along the side of the chamber 1/2 inch apart parallel to the electrodes. These wires are kept at the correct potential with the help of a resistor chain. Three independent sets of wires are used to smooth out the electric field. Measurements of the apparent curvature of tracks near the pole piece indicate that the coherent distortions in a 40 cm gap chamber are tolerable for most experiments beyond 1 inch from the lucite wall of the chamber.

Since the streamer development is a statistical process and the spark track starts as a string of streamers, it is not surprising that a spark track shows small differences in intensity and width, and small excursions around an average direction. As expected, the larger the magnitude and therefore the shorter the length of the applied pulse, the more pronounced are the differences in intensity and width.

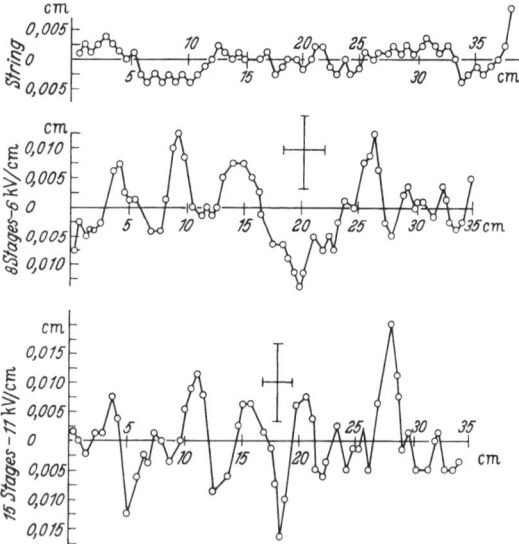

Fig. 4. The deviation of track "points" from the best fit circle. The abscissa represents the position along the track, and the ordinate the deviation from the best circle fit

To study the magnitude of this incoherent distortion and its possible influence on curvature determination, we have measured of the order of 50 "points" along cosmic rays tracks in a magnetic field with the Hermes measuring machine in its automatic track follower mode. A photo-electric-optical system automatically centers the track by averaging over a track segment 0.6 cm long. A "point" thus represents an average over this track length. Typical results are shown in Fig. 4.

The deviations shown in Fig. 4 represent deviations from the best circle fit of tracks observed in a 40 cm chamber. Results with and without magnetic field are very similar. The regions near the metal plates were not used. With a 11 kV/cm pulse a r. m. s. deviation amplitude d_i of ± 0.006 cm is observed. The characteristic track length l_t over which such a deviation occurs is 2.5 cm. With a 6 kV/cm pulse only the value of l_t is changed significantly to 4 cm.

The exact values of the above parameters depend on the condition of the chamber gas and the shape of the high voltage pulse. The values given above are typical of our set up under working conditions. It is indicated that even with the higher electric field the accuracy of the track fit is not appreciably increased by measuring "points" much more closely spaced than 2 cm, and that the typical error of a "point" should be ± 0.006 cm. This result applies to minimum ionization tracks. When two particles with different ionization losses traverse a chamber at the same time, the values of l_t could well be different for the two spark tracks.

2. Sensitive time. A particle traverses a chamber and triggers the high voltage. How long ago can an earlier particle have traversed the chamber and still appear as a spark track? This "sensitive time" is important for use of these chambers at accelerators. It is clearly a strong function of the gas purity since it depends on the probability of the attachment of ionization electron to heavy ion impurities.

It is easy to obtain an upper limit to the sensitive time by studying the efficiency for track formation as a function of the time delay between the passage of a particle. For gas under continuous purification and the chamber operating under stable conditions, tracks begin to have an "old look" after 2 μsec delay, but are seen with high efficiency for times up to 20 μsec. The addition of molecules with a high electron attachment probability reduces these times but, as expected, increases the incoherent track distortion at the same time.

In a recent experiment with 160 MeV protons, EISENSTEIN et al. [20] have studied the efficiency of spark track formation by protons arriving earlier than the one just before the H. V. pulse is applied. Less than 10% of the particles arriving more than 1 μsec before the H. V. pulse produce measurable tracks. For tracks of any kind the corresponding time is 4 μsec. Not surprisingly these more realistic values for the sensitive time are considerably shorter than the ones obtained by the delayed pulse method. For the streamer mode the delayed pulse mode should give a more correct value.

3. Stability of operation. When using lucite wall chambers, it is necessary to circulate the chamber gas through a purification system to keep the tracks uniform. With such a system we have observed $\cong$20,000 tracks over a period of 3 days without any apparent change in operating conditions or track appearance. Glass walls appear not to require any purification system.

4. Maximum permissible background. In a typical experiment, a tagged photon beam passes through a hydrogen target and then through a wide gap chamber located in a magnetic field. This chamber is designed to observe strongly interacting particles emitted in the forward direction from the target. Besides the desired particles, background Compton electrons and $e^+ e^-$ pairs are produced in the walls and the gas which produce in the chamber a certain density of positive ions after the more mobile electrons have been reattached. To reduce this background, we construct the electrodes out of 0.0015 inch diameter stainless steel wire mesh and use mylar windows.

The background charges present in the chamber will, if dense enough, prevent spark formation. What is the maximum photon beam that can be passed through a chamber located in a magnetic field without interfering with spark track formation? Unfortunately we have as yet not any information on this important point due to delay in magnet delivery. But we expect to make measurements in the near future.

A very useful development being carried out by KNASEL et al. [21] consists in putting a helium region separated by thin mylar partitions from the neon into that portion of the spark chamber through which passes the high intensity beam. This desensitizes this region.

5. Measurement of specific ionization. In the spark mode it is expected that the luminosity of a spark track depends on the ionizing power of the particle. This has in fact been observed and studied by LYUBIMOV et al. [22, 23] in pure Ne. In addition they find that if the Ne

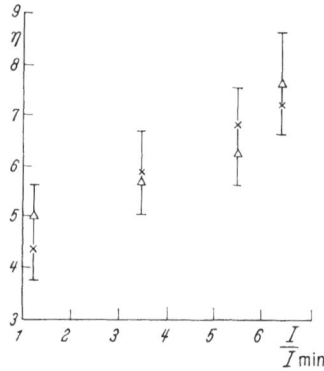

Fig. 5. Relation between specific ionization and luminous dot density obtained by ASATIANI et al. [25]

gas contains a small amount of propane and/or alcohol, the avalanche density is reduced sufficiently so that the track appears as a collection of "bunches". The "bunch" density increases with increasing ionizing power. While the luminosity of a given track can depend on many other factors besides the specific ionization, the bunch density appears to depend mainly on the particle ionization. For particle events in which the angle of the tracks with respect to electric field does not vary much, the luminosity seems to be a promising tool for particle identification. Where high spatial resolution is not required, the bunch density in dirty neon appears to be a useful feature of the spark track.

Because the speed of avalanche development depends on the number of initiating electrons, the length of streamers and therefore their luminosity should depend on the ionizing power of the initiating particle. This effect was first studied theoretically and experimentally by CHIKOVANI [24] who also suggested that the luminous dot density could be another measure of ionizing power.

An interesting investigation by ASATIANI et al. has been reported by ALIKHANYAN [25]. They used protons with energies of 1) 660 MeV, 2) 100 MeV, 3) 40 MeV, 4) 20 MeV and photographed a) parallel (film speed 1300, objective $f/2$ and $f/4$ and b) perpendicular (film speed 3000, objective $f/1.5$) to the electric field direction. The effect of ionizing power on the tracks can be seen by eye. Some of their preliminary results are summarized in Fig. 5 where is shown the dependence of the number n of luminous dots on specific ionization. The effect is there, but is not large. It seems desirable to continue this work with different operating conditions to see whether it is not possible to obtain a larger effect.

E. Conclusions

The large amount of physics presented at this conference obtained with H_2 bubble chambers and spark chambers shows the usefulness of these devices at electron accelerators. The future use of enriched, polarized or tagged beams with bubble chambers, and of the new type of track spark chambers promises to bring additional understanding to the field of photon and electron interactions.

References

[1] Research contributions, session I
[2] BINNIE, D. M.: Nucl. Instr. and Methods 10, 212 (1961)
[3] BALLAM, J., and Z. GUIROGOSSIAN: Proc. International Conf. on Instrumentation, Dubna 1964, to be published
[4] TSAI, Y. S.: Phys. Rev. 137 B, 730 (1965)
[5] BARBIELLINI, G., G. BOLOGNA, G. DIAMBRINI, and G. P. MURTAS: Phys. Rev. Letters 9, 396 (1962), contains references to earlier papers
[6] MILBURN, R. H.: Phys. Rev. Letters 10, 75 (1963)
[7] AZUTYUNIAN, F. R., and V. A. TUMANIAN: Phys. Letters 4, 176 (1963)
[8] — I. I. GOLDMAN, and V. A. TUMANIAN: Zurn. eksper. teor. Fiz. 45, 312 (1963); translated in Soviet Physics JETP 18, 218 (1964)
[9] BEMPORAD, C., M. FOTINO, R. H. MILBURN, and N. TANAKA: CEA Report-1014 (1965)
[10] MILBURN, R. H.: Report SLAC-41 (1964)
[11] — Session 3 A, Paper 9
[12] TAYLOR, R. E., and R. F. MOZLEY: Phys. Rev. 117, 835 (1960)
[13] SMITH, R. C., and R. F. MOZLEY: Phys. Rev. 130, 2429 (1963)
[14] CALDWELL, D. O., K. HEINLOTH, J. P. DOWD, and T. R. SHERWOOD: Paper 6 Session 1 B

[15] STRAUCH, K.: Proceedings of the Purdue Conference on Instrumentation in High Energy Physics, to be published (1965)

[16] CHIKOVANI, G. E., V. A. MIKHAILOV, and V. N. ROINISHVILI: Phys. Letters 6, 254 (1963), and Zurn. eksper. teor. Fiz. 45, 818 (1963)

[17] DOLGOSHEIN, B. A., and B. I. LUCHKOV: Zurn. eksper. teor. Fiz. 46, 1 (1964)

[18] ALIKHANYAN, A. I., T. L. ASATIANI, E. M. MATEVOSIAN, and R. D. SHARKHA-TUNIAN: Phys. Letters 4, 295 (1963)

[19] GARRON, J. P., D. GROSSMAN, and K. STRAUCH: Rev. Sci. Inst. 36, 264 (1965)

[20] EISENSTEIN, R., D. GROSSMAN, R. SAH, and K. STRAUCH: to be published in Nucl. Instr. and Meth.

[21] KNASEL T. M., J. K WALKER, and M. WONG: Rev. Sci. Instr. 36, 270 (1965)

[22] LYUBIMOV, V. A., Y. V. GALAKTIONOV, F. A. PAVLOVSKY, F. A. YETCH, and I. V. SIDOROV: ITEF Preprint 1964

[23] —, and F. A. PAVLOVSKY: Nucl. Instr. and Meth. 27, 342 (1964)

[24] CHIKOVANI, G. E., V. N. ROINISHVILI, and V. A. MIKHAILOV: Nucl. Inst. and Meth. 29, 261 (1964)

[25] ALIKHANYAN, A. I.: Work reported. Loeb Lectures, Harvard University 1965

Prof. Dr. K. STRAUCH

Lyman Laboratory of Physics
Harvard University of Cambridge
Cambridge/Mass.

SPRINGER TRACTS

IN MODERN PHYSICS

Ergebnisse
der exakten Natur-
wissenschaften

Volume **39**

Editor: G. Höhler
Editorial Board:
S. Flügge, F. Hund
E. A. Niekisch

Reprint:

L. van Hove
Theory of Strong Interactions of Elementary Particles in the
GeV Region

Springer-Verlag Berlin Heidelberg GmbH 1965

SPRINGER-VERLAG
BERLIN · HEIDELBERG · NEW YORK

Springer Tracts in Modern Physics
Ergebnisse der exakten Naturwissenschaften

SPRINGER TRACTS
IN MODERN PHYSICS

Ergebnisse
der exakten Natur-
wissenschaften

Volume **39**

Editor: G. Höhler
Editorial Board:
S. Flügge, F. Hund
E. A. Niekisch

Reprint:

R. Hofstadter et al.
Recent High Energy Electron Investigations at Stanford University

Springer-Verlag Berlin Heidelberg GmbH 1965

SPRINGER TRACTS
IN MODERN PHYSICS

Ergebnisse
der exakten Natur-
wissenschaften

Volume **39**

Editor: G. Höhler
Editorial Board:
S. Flügge, F. Hund
E. A. Niekisch

Reprint:

R. Wilson
Review of Nucleon Form Factors

Springer-Verlag Berlin Heidelberg GmbH 1965

SPRINGER TRACTS

IN MODERN PHYSICS

Ergebnisse
der exakten Natur-
wissenschaften

Volume **39**

Editor: G. Höhler
Editorial Board:
S. Flügge, F. Hund
E. A. Niekisch

Reprint:

G. Höhler
Special Models and Predictions for Pion Photoproduction (Low Energies)

Springer-Verlag Berlin Heidelberg GmbH 1965

SPRINGER TRACTS

IN MODERN PHYSICS

Ergebnisse
der exakten Natur-
wissenschaften

Volume **39**

Editor: G. Höhler
Editorial Board:
S. Flügge, F. Hund
E. A. Niekisch

Reprint:

S. D. Drell
Special Models and Predictions for Photoproduction above 1 GeV

Springer-Verlag Berlin Heidelberg GmbH 1965

SPRINGER TRACTS
IN MODERN PHYSICS

Ergebnisse
der exakten Natur-
wissenschaften

Volume **39**

Editor: G. Höhler
Editorial Board:
S. Flügge, F. Hund
E. A. Niekisch

Reprint:

L. S. Osborne
Photoproduction of Mesons in the GeV Range

Springer-Verlag Berlin Heidelberg GmbH 1965

SPRINGER TRACTS

IN MODERN PHYSICS

Ergebnisse
der exakten Natur-
wissenschaften

Volume **39**

Editor: G. Höhler
Editorial Board:
S. Flügge, F. Hund
E. A. Niekisch

Reprint:

R. Gatto
Theoretical Aspects of Colliding Beam Experiments

Springer-Verlag Berlin Heidelberg GmbH 1965

SPRINGER TRACTS

IN MODERN PHYSICS

Ergebnisse
der exakten Natur-
wissenschaften

Volume **39**

Editor: G. Höhler
Editorial Board:
S. Flügge, F. Hund
E. A. Niekisch

Reprint:

W. K. H. Panofsky
Experimental Techniques

Springer-Verlag Berlin Heidelberg GmbH 1965

SPRINGER TRACTS
IN MODERN PHYSICS

Ergebnisse
der exakten Natur-
wissenschaften

Volume **39**

Editor: G. Höhler
Editorial Board:
S. Flügge, F. Hund
E. A. Niekisch

Reprint:

K. Strauch
The Use of Bubble Chambers and Spark Chambers at
Electron Accelerators

Springer-Verlag Berlin Heidelberg GmbH 1965